# Multilevel Interconnect Technology

## Intel/McGraw-Hill Series

BUNZEL and MORRIS • *Multimedia Applications Development: Using Indeo™ Video and DVI® Technology*

EDELHART • *Intel's Official Guide to 386 Computing*

GUREWICH • *Communication Systems: Practical Guide to Intel's Connectivity Design*

INTEL • *Intel386 Family Binary Compatibility Specifications 2*

LUTHER • *Digital Video in the PC Environment*

MARGULIS • *i860 Microprocessor Architecture*

RAGSDALE • *Parallel Programming*

YUEN • *Intel's SL Architecture: Designing Portable Applications*

# Multilevel Interconnect Technology

Gopal K. Rao

**McGraw-Hill, Inc.**

New York San Francisco Washington, D.C. Auckland Bogotá
Caracas Lisbon London Madrid Mexico City Milan
Montreal New Delhi San Juan Singapore
Sydney Tokyo Toronto

**Library of Congress Cataloging-in-Publication Data**

Rao, Gopal K.
Multilevel interconnect technology / Gopal K. Rao.
p. cm.—(Intel/McGraw-Hill series)
Includes bibliographical references and index.
ISBN 0-07-051224-8
1. Integrated circuits—Very large scale integration—Design and construction. 2. Electronic circuit design. 3. Integrated circuits—Very large scale integration—Reliability. 4. Production engineering. I. Title. II. Series.
TK7874.R35 1993
621.39′5—dc20 93-20583
CIP



1 2 3 4 5 6 7 8 9 0 DOC/DOC 9 9 8 7 6 5 4 3

ISBN 0-07-051224-8

*The sponsoring editor for this book was Daniel A. Gonneau, the editing supervisor was Stephen M. Smith, and the production supervisor was Pamela A. Pelton. It was set in Century Schoolbook by McGraw-Hill's Professional Book Group composition unit.*

*Printed and bound by R. R. Donnelley & Sons Company.*

*To my parents*

# Contents

# Foreword

Multilevel interconnect technology is a very important and integral part of the design and fabrication of submicron VLSI/ULSI integrated circuits. It has become a crucial area for extensive research and development because it is needed for enhancing the performance of advanced integrated circuits. The density and functionality of today's microprocessors are limited by our ability to provide dense multilevel interconnections.

This book, *Multilevel Interconnect Technology*, takes a systematic approach to the description of on-chip interconnect technology and its implementation in high-volume manufacturing. Written from the perspective of the practitioner, the book contains extremely useful information on the design and fabrication of interconnect systems in high-performance integrated circuits. This book serves as a basic reference for this complex field and should be of interest to a wide range of people, including designers, process engineers, reliability engineers, and graduate students.

*Dr. Gerry Parker*
*Senior Vice President and General Manager*
*Technology and Manufacturing Group*
*Intel Corporation*

# Preface

This book, *Multilevel Interconnect Technology*, deals exclusively with the subject of fabricating submicron devices using multiple levels of metal interconnects. This discussion is important since multilevel interconnect technology is the technology employed in the fabrication of advanced integrated circuits.

What is multilevel interconnect technology? It is the technology that consists of the design and processing steps that are needed to connect the active areas on a device. As device geometries shrink and more and more transistors are packed in a single chip, interconnections between active areas can only be accomplished by stacking one level of metal interconnects over the other in order to save silicon real estate.

Successful transfer of product designs to silicon and subsequent high-volume fabrication allows an IC manufacturer to gain competitive advantage in the marketplace. Fabrication of devices may be broken into two parts which are common to all products: formation of transistors and connecting the transistors. The former has been studied since the invention of transistors; the latter, especially the use of multiple levels of metallization, is a newer technology that is still being characterized and studied. Therefore, multilevel interconnect technology may very well be the gating item for high-volume submicron IC fabrication.

This book is intended for use as a reference guide for engineers and graduate students involved with advanced IC fabrication. The book presents an integrated overview of multilevel interconnect technology so that readers can understand and appreciate the causes and effects of various building blocks in the fabrication of ICs, from design to reliability performance. Cause and effect diagrams provide an excellent overview of all major parameters that affect a particular problem or process. Whenever possible, tables and illustrations—generic and specific—aid in the understanding of issues.

This book is organized into four chapters that provide a qualitative and coherent perspective of the issues pertaining to multilevel interconnects.

Chapter 1 examines the changing technology scenario and its impact on interconnect technology. This chapter also deals with design issues specific to multilevel interconnects. The focus is on the impact on RC delays and layout due to scaling.

Chapter 2 discusses process development issues pertaining to the transfer of multilevel interconnect designs to silicon. The focus is on the methodology for process flow definition, with emphasis on materials and their interactions.

Chapter 3 deals with high-volume IC manufacturing. The focus is primarily on cycle time reduction, yield analysis, and defect reduction. High line and die yields are a must to reduce wafer costs and to introduce products into the marketplace quickly.

Chapter 4 explores the reliability issues of interconnects. As the interconnect levels and density increase, the challenge is to fabricate highly reliable ICs by designing in reliability during all phases of IC design and fabrication.

*Gopal K. Rao*

# Acknowledgments

The task of writing a book has been very pleasant and exciting, primarily because of the challenges and opportunities that come with such an endeavor. It is indeed a privilege to be associated with Intel and McGraw-Hill and share the front cover of this book with them. I am very grateful to Intel for providing me the environment, experience, and opportunity to write this book. I am also thankful to McGraw-Hill for publishing this book; specifically, I wish to thank Daniel Gonneau and Stephen Smith, my editors, for their parts in the publication process.

Special thanks are due to Mark Bohr, Janet Brownstone, John Carruthers, Dave Fraser, Scot Ruska, Quat Vu, and Mike Wegener for undertaking the time-consuming task of reviewing the manuscript.

I would like to thank Lloyd Gravley and Scott Luebke for the artwork and Carole Bigelow for the colorful cover design and slide preparation. Thanks are also extended to Rob Gordon, Ann Nelson, and Mary Banwart. I wish to thank Anita Lee, Heather O'Danniel, and Sharon Lundquist for their help with literature searches.

A big thank you goes to my wife, Lalita, and my children for cheering me on to the finish line through constant encouragement and support. I am also grateful to my brothers and sisters, especially my brother Dr. K. N. Rao and my late sister P. Boray, for their encouragement and support.

Finally, the publication of this book is genuinely a reflection of the goodwill and support of a lot of people—ingredients that are essential for all authors. I am deeply indebted to all those who have made it possible for me to write this book. I sincerely hope that this book serves as a valuable aid in addressing the technological challenges that confront us in the fabricatin of integrated circuits.

## ABOUT THE AUTHOR

Gopal K. Rao is a senior process integration engineer at Intel Corporation in Rio Rancho, New Mexico. Before joining Intel, he worked at Philips Research Laboratories, Texas Instruments, and DCM Data Products. Mr. Rao has more than 13 years of experience in the semiconductor and computer industries, and has held various positions in engineering, management, and marketing. He has also authored several technical papers on various aspects of multilevel interconnect technology, and holds one patent.

Chapter

# 1

# Design

## 1.1 Introduction

Multilevel interconnect design has become very complicated as integrated circuits (ICs) have increased in complexity and size. The challenge in interconnect design is to maximize circuit performance. This may be accomplished by minimizing propagation delays and optimizing interconnect line layout on the die. The two requirements must be satisfied simultaneously without compromising either one.

Propagation delays due to parasitic capacitances from interconnects are one of the main causes for compromising speed performance in advanced ICs as interconnect dimensions are scaled down. In addition, increases in die size make interconnect lengths very long, which means higher interconnect resistance also contributes to propagation delays due to an increase in RC time constant. This is a product of the interconnect resistance and capacitance. Innovations in design techniques and the use of electrically superior interconnect and dielectric materials are needed to reduce propagation delays.

Interconnect layout reflects a designer's skill in manipulating available die area by using design rules and packing density to connect all necessary components on a die. Predesigned and tested circuit building blocks and powerful computer-aided design (CAD) tools help reduce design effort and expedite turnaround of designs for manufacturing. Algorithms are written to determine the best routing scheme for interblock and intrablock connections. Interconnect layout is done within the framework of design rules formulated for a given device. These design rules ensure products are functional within the limits specified. The product and process requirements must be balanced when defining design rules for products. Feedback from process development, manufacturing, and reliability groups is constantly sought to ensure design rules are not violated. If design rules are violated due

**TABLE 1.1 Four Major Design Thrusts**

| Design for functionality | Design for testability | Design for manufacturability | Design for marketability |
|---|---|---|---|
| Design rule violation | Test methodology | Design rule violation | Product specifications |
| Layout connectivity errors | Layout verification | Layout density and defect density | Product demand and supply |
| Circuit modeling and verification | Circuit simulation and verification | Product demand and supply | Benchmarking and performance |
| Yield | Benchmarking | Yield | Design debugging |
| Performance | Design debugging | Reliability | Reliability |

to process margins, then the implications need to be carefully evaluated. This leads into the subject of design for manufacturability.

The objective is to develop designs that provide desired device applications without reducing their degree of manufacturability. Obviously, it is more profitable to develop designs that are conducive to high-volume manufacturing. Designs must also be testable; testability of designs at the development stage and the ability to debug design problems during the product sampling stage are a must. The turnaround time required to fix design problems prior to high-volume manufacturing is critical for success.

For a clearer understanding of the design issues, the entire IC design program may be divided into four groups. Any IC design must cater to all of the requirements in each group. As shown in Table 1.1, there is an intentional overlap of issues between each of these groups. Let us examine how, for example, interconnect layout can influence all of the four groups. If there are layout connectivity errors, then the device will not function. To test if an interconnect layout scheme yields functional circuits, layout extraction and verification methodologies must be developed. An interconnect layout scheme can also degrade the die yield and reliability during manufacturing as it may be more susceptible to higher defect density due to placement and routing of interconnects.

A similar interrelationship can be established for various other design parameters. For maintaining a good track record of design wins, a strong development effort is needed in each of the four areas. This chapter focuses on interconnect design issues.

## 1.2 Technology Scenario

The following perspectives on the technology are interrelated and influence each other either directly or indirectly. Therefore, the role and responsibility of each is briefly examined to give the reader a flavor of the issues facing the development and introduction of new products and the pace needed to maintain and enhance the competitive edge.

### 1.2.1 System technology

Innovations in design, process, and manufacturing are propelled by demands for new consumer, military, and industrial products. The speed of introduction and performance standards for new products present various technological challenges in the fabrication and packaging of integrated circuits. Widespread use of computers and peripheral devices is providing an impetus for the development of IC technology. The pace of new product introductions has been rapid with the miniaturization of computing power. At present, computers are available in various forms, such as desktop, laptop, notebook, and even handheld. Increased demand for enhanced graphics capabilities and communication have made computers powerful tools in data and information management. In order to support these advances in computers, new IC products with increased memory and logic functions are becoming necessary.

### 1.2.2 Device technology

On the central processing unit (CPU) front, new microprocessors with high clock frequencies and functions are being introduced. For example, reduced instruction set computing (RISC) architectures introduce complexities into interconnect and packaging systems where the input/output (I/O) timing issues at high clock frequencies need to be addressed. On the data storage front, dynamic random-access memory (DRAM), static random-access memory (SRAM), and electrically programmable read-only memory (EPROM) devices with increasing memory capacity and decreasing data access times are also keeping pace with microprocessor development. The rate of introduction of these IC components has increased over the last few years as competition between manufacturers intensifies. Early introduction of new IC products enables system manufacturers to evaluate the capabilities of these IC components in their own systems. This gives system manufacturers an early start to market their systems with new features to satisfy market demands. To meet the speed/performance challenge, complementary metal-oxide semiconductor (CMOS) device technology is transitioning into newer technologies, such as bipolar CMOS (BiCMOS). This technology is a combination of CMOS and bipolar device technologies.

### 1.2.3 Design technology

Design technology development is faced with two challenges. The first is designing circuits that offer a broader set of applications and are compatible with each new generation. Compatibility is the key to ensuring the product line is sustained with each successive design modification and improvement. The design technology has to support new directions in device technologies, e.g., BiCMOS. The second challenge in design technology is in the layout of interconnects in the IC as the device dimensions are scaled down to enhance speed and increase packing density. To accommodate both these changes in as little area as possible, the use of multiple levels of interconnects has become necessary. Interconnect layout methodologies, to minimize parasitics and optimize timing, design rule specifications, and ease of circuit testability are some of the salient areas of design development, as there is demand for circuits with high clock frequencies and low power consumption.

### 1.2.4 Process technology

Process technology is progressing on a parallel path to design technology development—enhancing transistor characteristics and developing newer materials for dielectrics and interconnects. The main challenge is developing processes that can successfully transfer designs into silicon (Si) in a manufacturable way. The pace of transistor innovation and integration has been rapid, however, and development of interconnect technology has not kept up with it. New generations of designs impose even more stringent requirements on interconnect processes.

### 1.2.5 Manufacturing technology

Die size increases and downward scaling of device and interconnect dimensions have made defect reduction and yield enhancement major concerns during high-volume manufacturing of integrated circuits. Can new designs be developed that make the devices defect tolerant? One of the common bottlenecks in manufacturing is the introduction of design iterations that are done in response to design bugs. How can design iterations be optimized and minimized? These are some of the design-related questions that may slow down high-volume production of ICs. A key challenge is developing designs for ease of manufacturability and testability.

### 1.2.6 Reliability

A successful product design incorporates test structures that can evaluate various reliability issues. Designs and processes are being developed to make ICs more robust and increase their operating life in the field. Specifically, interconnect reliability issues have become the focal point for reliability enhancements.

### 1.2.7 Technology drivers

The semiconductor industry has relied on technology drivers to hasten the pace of development in various aspects of IC design, processing, and manufacturing. Memory devices have traditionally been used to enhance process and high-volume manufacturing technology, and application-specific integrated circuits (ASICs) have been used to exploit new techniques in designing circuits, packaging, and flexible manufacturing. Figure 1.1 is an example of both process and technology drivers in the IC industry and also presents activities that are influenced by these drivers.[1] For example, process technology drivers such as SRAMs possess regular arrays which can be used to debug process related problems relatively easily.

### 1.2.8 Product life cycle

It is important to comprehend product life cycles to appreciate the time available to design, manufacture, and market new products in order to stay competitive in the marketplace. All those involved in semiconductor technology need to be sensitive to this to make their design, process, and production technologies aligned to achieving reliable devices at the lowest possible cost. The life cycle of a product con-

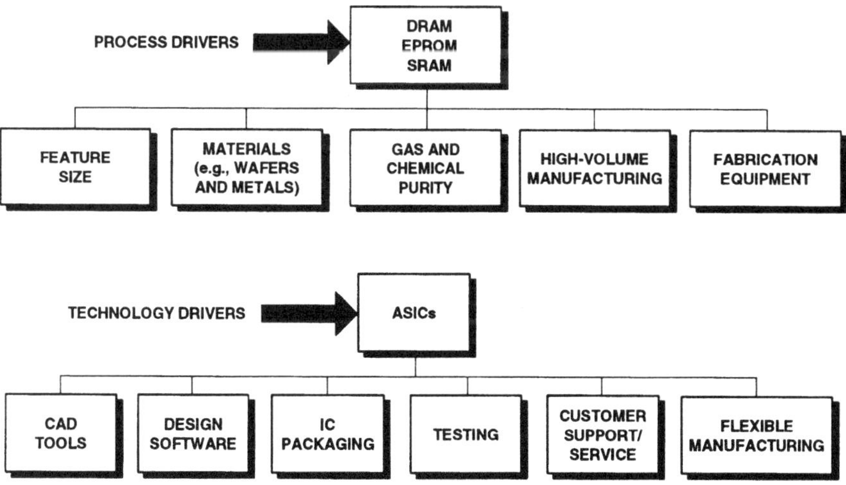


**Figure 1.1** Process and technology drivers in the IC industry.[1] (*Courtesy of Integrated Circuit Engineering Corporation.*)

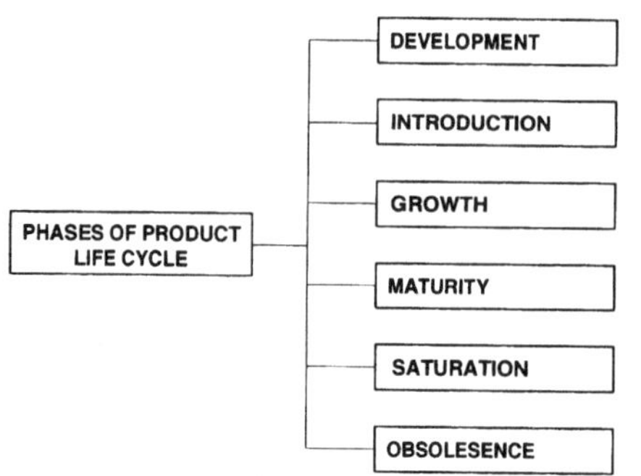


**Figure 1.2** Six phases in a product life cycle.[1]

sists of six phases. These phases take a product from being proprietary to commodity. Figure 1.2 shows the six phases of a product life cycle.[1] With current levels of competition in the market place, the time between product development and introduction is becoming smaller. The key to adapting to this change is to reduce, but not compromise, the design and process development cycle. Learning from previous technologies and improving on them is one way to approach this problem.

Timing is key to success. Introduction of new products when competition is relatively soft is beneficial in capturing market share and getting a head start in debugging design problems and fine tuning manufacturing processes for high volume and yield. As a product matures, the competition intensifies as more manufacturers enter the marketplace, therefore, product price becomes a major issue. The key to success in this time frame is low manufacturing costs. Old product demand starts to saturate as new products are developed and introduced. Finally, it is no longer cost-effective for either the manufacturer or the customer to continue using old products and migration to new products with superior features becomes inevitable.

### 1.2.9 New technology challenge

As described above, the process technology drivers have been the memory chips. The pace of process development has been very fast in order to keep up with the ever increasing demands for more memory. Therefore, design and process technologies had to migrate to submicron geometries. It is important to be aware of possible changes in the direction of technology and understand how they can impact multilevel interconnect designs and processes.

Integrated circuits may be divided into three areas: active regions, isolation regions, and interconnects. Even though the primary focus of this book is on interconnects, a thorough understanding of the changes that may take place in the development of active and isolation areas of a transistor is essential. The changes have to be integrated with developments in interconnect technology; interconnect technology can not be designed and developed in isolation.

CMOS technology has been the main tool in fabricating memory and microprocessor circuits. As CMOS technology migrates to submicron dimensions we are faced with some inherent limitations due to scaling. The impact of scaling on transistor properties is described in Section 1.3. Although device dimensions have been scaled down, the power supply has not due to system requirements. When the operating voltages are not scaled, the electric fields increase.This causes reliability problems due to hot carriers. Migration to lower operating voltages (3.3 V) will improve this to some extent but issues associated with noise margin and drive capability will still be outstanding. As interconnect and capacitive loads do not scale linearly, the drive current must be increased to gain further speed performance.

To overcome some of these problems, one solution has been to merge bipolar and CMOS technologies to form BiCMOS devices.[2,3] BiCMOS devices can deliver a significant increase in performance with minimal increase in process complexity. Figure 1.3 compares speed and power performance for technologies between CMOS and emitter-coupled logic (ECL).[2] CMOS has had the advantage of low power consumption; adding bipolar circuitry enhances the speed but also increases the power consumption. The benefit of bipolar circuits is the capability to use higher on-chip and off-chip current drive capability. Although new materials and improved design and layout of interconnects can decrease RC time delays, the use of improved circuit techniques like BiCMOS can help decrease the power dissipation at driver circuits. Alternating current (AC) power dissipation is a problem in CMOS circuits and this occurs at circuits (driver and buffer) that drive large capacitive loads.

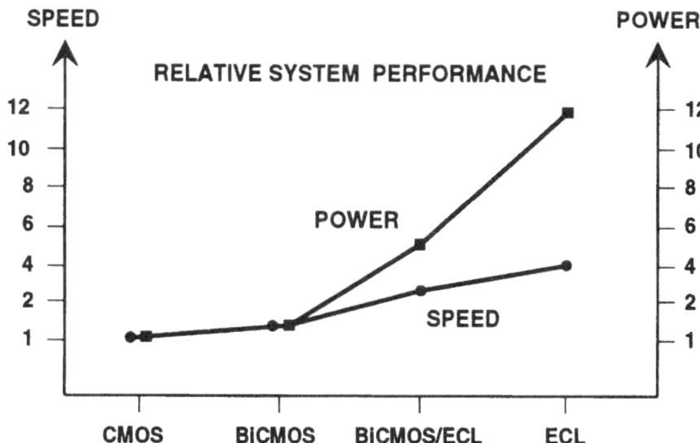


**Figure 1.3** Performance comparison of CMOS, BiCMOS, and ECL technologies.[2] (*Courtesy of Semiconductor International.*)

## 1.3 Scaling

The challenge is to pack more and more transistors on a single chip in order to offer customers a greater set of device features and applications. Downward scaling of feature sizes allows an increase in circuit speed and packing density. This also decreases power dissipation. There are some negative effects of downward scaling on transistor, interconnect, and reliability parameters; however, these arise when the dimensions are scaled down and operating voltages are not, leading to various short-channel effects. In CMOS circuits, scaling can form parasitic bipolar transistors that may turn on and cause latch-up, which damages the circuit.

As interconnect dimensions decrease and clock frequencies increase, RC time delays become the predominant barrier to achieving high circuit speeds. There are several process issues that arise from scaling interconnect dimensions. For example, as metal pitches decrease, dielectric deposition and planarization become more difficult. Interconnect technology may be the bottleneck in realizing potential improvements in device speed, packing density, and reliability that stem from downward scaling of device parameters. Processing problems associated with the definition of the interconnect under a scaled environment are increasing as the device dimensions migrate to sub-half micron regime. From a design perspective, the scaling of interconnects presents a challenge to manage resistance and capacitance issues to maintain specified circuit speed due to RC delays varying with geometry. In addition, as the dimensions go below 1 μm, the current density and resistivity of the metal films pose severe limitations in achieving the desired speed performance specification for devices.

When the supply voltage is unchanged while the other parameters are scaled down, electric fields in the gate region increase. This leads to hot carrier degradation and oxide breakdown. The reliability of the interconnects is another major side effect of scaling down the dimensions. Increased current density in narrow metal lines and a higher contact resistance make Al interconnects more susceptible to electromigration (EM), stress voiding, and cracking failures.

### 1.3.1 Transistor scaling[4–9]

To enhance device speed and increase packing density, transistor geometries need to be reduced. The primary objective during this reduction process is to ensure that the electric field patterns in the smaller device are identical to those in the larger device. That is, the physical and electrical parameters of interconnects should be scaled down appropriately (scaling results in increase in electric fields) to keep the direction and magnitude of the electric fields constant. The outcome of this achievement is an increase in the speed of a device.

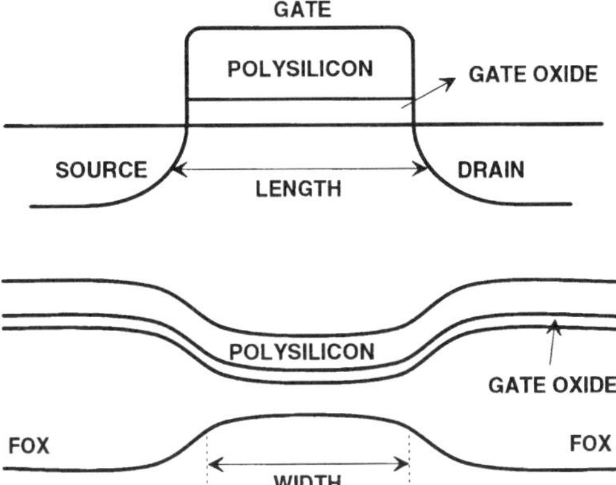


**Figure 1.4** Schematic diagram of a field-effect transistor.

Let us examine the issues associated with scaling by using a schematic diagram of a field-effect transistor, as shown in Figure 1.4. The key physical dimensions are shown in this diagram. To make the transistor occupy a smaller area, the channel length and width can be reduced, leading to increased device speed. Table 1.2 presents various parameters and their scaling factor for ideal scaling. The positive effects of scaling are clearly evident from Table 1.2. However, there are some negative side effects of scaling due to nonscaling of operating voltages; these are highlighted in Figure 1.5.

Under ideal scaling (constant electric field) conditions[7], all geometries and operating voltages are scaled down by a scaling factor $1/S$ while the substrate doping is increased by $S$, where $S$ is greater than 1. As a result of this scaling, the electric fields remain approximately unchanged and consequently velocity saturation and carrier heating

**TABLE 1.2 Ideal MOS Transistor Scaling**

| Parameter | Scaling factor |
|---|---|
| Lateral dimensions $[W, L]$ | $1/S$ |
| Vertical dimensions $[t_{gox}, X_j]$ | $1/S$ |
| Substrate doping $[N_{sub}]$ | $S$ |
| Voltages | $1/S$ |
| Current | $1/S$ |
| Gate capacitance | $1/S$ |
| Gate delay | $1/S$ |
| Power dissipation/gate | $1/S^2$ |
| Power density | 1 |
| Power-delay product | $1/S^3$ |

Abbreviations: $t_{gox}$ = gate oxide thickness; $X_j$ = junction depth.

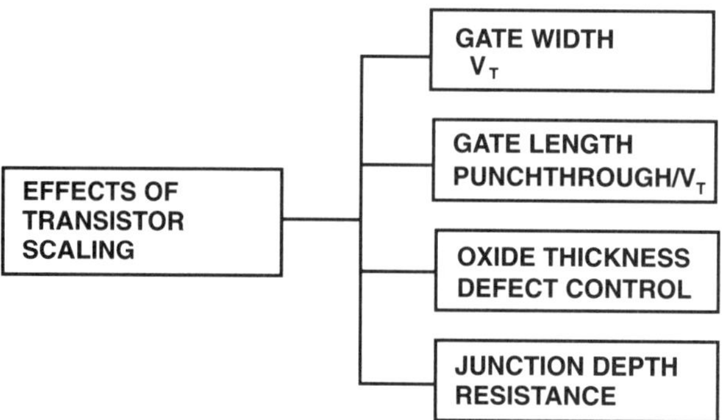


**Figure 1.5** Effects of transistor scaling.

are prevented.[4,6] With ideal scaling, devices get faster, power dissipation and gate delay are reduced, and packing density improves.

In spite of some obvious advantages to voltage scaling, the operating voltages have remained unchanged at 5 V ± 10 percent for a long time to satisfy transistor transistor logic (TTL) compatibility and ensure continued usage of existing ICs without installing multiple power supply sources in various systems. For submicron devices, specifically under 0.7 μm, the industry wide standard for operating voltage is 3.3 V ± 10 percent. Figure 1.6 highlights some of the key problems of operating voltages being held constant while other parameters are scaled down.

**Lateral dimensions reduced/voltage constant.** Both the gate length and width modulate the threshold voltage of a transistor in addition to the drain voltage. As the channel length gets shorter, control of threshold voltage ($V_T$) becomes more difficult due to drain-induced barrier lowering.[8] As the channel length decreases, the threshold volt-

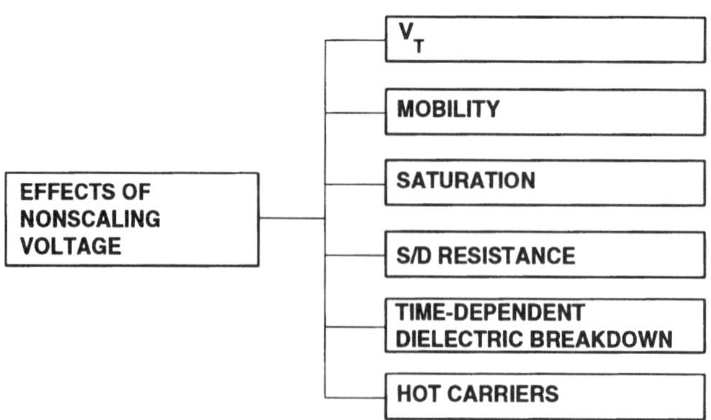


**Figure 1.6** Effects of nonscaling operating voltages.

age also starts to decrease. When the gate width is reduced, the relationship between $V_T$ and width has the opposite effect of length reduction on $V_T$. As the width decreases, the threshold voltage increases after a certain value. The threshold voltage is dependent on the gate width because of electrical field spreading outside the gate region due to lateral encroachment of dopants. This changes the amount of depletion charge associated in with the gate region.

**Length reduced/voltage constant.** As the channel length is reduced under constant voltage conditions, the electric fields along the channel increase. The drift velocity of carriers is directly proportional to the applied electric field and continuously increases until it reaches a saturation point. The saturation velocity between n-type and p-type silicon varies due to differences in carrier mobility. Degradation in device transconductance is observed as the carrier velocity saturates. This impacts the drive capability of transistors.

**Lateral and vertical dimensions reduced/voltage constant.** Under these conditions, the mobility of the carriers is also affected. As the channel electric fields increase, the mobility is reduced for both p-channel and n-channel devices and degrades the device performance.

**Thickness reduced/voltage constant.** Gate oxide thicknesses are being reduced significantly, and this is not accompanied by the requisite scaling of the supply voltage. This leads to higher electric fields over the gate, causing gate oxide breakdown. Gate oxide failures (breakdown and wearout) are a reliability hazard as they have a time-dependent nature.

**Junction depth reduced/voltage constant.** When the junction depths are scaled down, the source/drain resistances increase, resulting in degradation in the transconductance of the device. Junction depth scaling has significant implications on multilevel interconnect

**TABLE 1.3 Implications of Junction Depth Scaling**

| Cause | Effect | Reaction |
|---|---|---|
| High source and drain resistance | Leads to high contact resistance, EM degradation, and reduced device speed | 1. Silicides on junctions and on polysilicon interconnects to lower the resistance<br>2. High selectivity to silicide films needed during contact etch process as silicide films are an etch stop |
| Junction spiking | Junction breakdown due to Al spiking | 1. Barrier layers between Al interconnects and junction. Saturate Al with Si. |

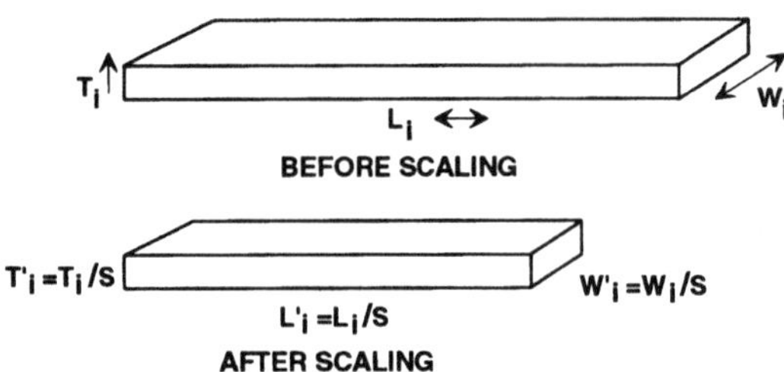


**Figure 1.7** Schematic diagram illustrating interconnect scaling.

processes, which are summarized in Table 1.3. Junction depth scaling also impacts contact resistance and EM. Managing contact etch processes to achieve desired profiles and selectivity to underlying films (silicide layers) is a major aspect of controlling contact resistance of the devices (this concept is further elaborated in Chapter 2).

### 1.3.2 Interconnect scaling[6,7,10–16]

The concept of scaling may be visualized by the simple illustration given in Figure 1.7. The original interconnect dimensions are reduced by a scaling factor, $S$, and the die size scaling factor is $S_c$. It is assumed that as the die size increases the interconnect lengths also increase. Both $S$ and $S_c$ are considered to be greater than 1. For example, in Figure 1.7, the new scaled interconnect length, $L_i'$, and width, $W_i'$, may be expressed as follows:

$$L_i' = L_i/S \tag{1.1}$$

$$W_i' = W_i/S \tag{1.2}$$

Let us use this basic concept to examine the impact of scaling on interconnect resistance and capacitance.

**Interconnect resistance.** Interconnections serve several functions on the chip and may be divided into two groups. First, local interconnects are used to wire a group of both active and passive elements in close proximity. Generally, these are thin and densely packed metal lines. Second, global interconnections are used to connect the various circuit elements. It is these interconnections that limit the speed of the device. The design rules for each of these interconnects are different owing to their diverse applications. The interconnections, besides carrying current to the active areas, also serve as an interface to the package metallization. The topmost metal lines that carry the bulk of the power are wider and thicker, thereby reducing the resistance. The global interconnect pitch and length are determined by the cell size and the layout required for each cell. Interconnect resistance increases as the length increases and cross-sectional area is reduced. In addition to scaling, the cross-section of metal lines may be reduced as a result of process variations, causing interconnect resistance to increase. For example, if a metal line has poor step coverage (described in Chapter 2) the cross-sectional area will be reduced, offering a higher resistance to the current flow. In some cases, poor step coverage of metal over steps may lead to EM failures. The line resistance, $R_i$, may be expressed as follows for local and global interconnects, respectively:

For local interconnects $$R_i = \rho \frac{L_i}{W_i T_i} \tag{1.3}$$

where $\rho$ = resistivity of interconnect
$L_i$ = interconnect length
$W_i$ = interconnect width
$T_i$ = interconnect thickness

For global interconnects $$R_i = \rho \frac{L_g}{W_i T_i} \tag{1.4}$$

where $\rho$ = resistivity of interconnect
$L_g$ = average global interconnect length
$W_i$ = interconnect width
$T_i$ = interconnect thickness

The average global line length may be expressed as follows:

$$L_g = A^{1/2}/2 \tag{1.5}$$

where $A$ = chip area ($\mu m^2$)

The scaling factor, $S_c$, for global interconnects is considered to be equivalent to its length. When all the dimensions are scaled down by the scaling factor, $S$, the scaled line resistance, $R_i'$, may be expressed in the following way:

For local interconnects $$R_i' = SR_i \tag{1.6}$$

For global interconnects $$R_i' = S^2 S_c R_i \tag{1.7}$$

Figure 1.8 illustrates how the interconnect resistance varies as the line width is increased from 0.5 to 5.0 μm, and highlights the relative contributions of line width and thickness to interconnect resistance.[12] To increase transistor packing density, feature sizes are scaled down and the die size is increased. As the die size increases, so does the length of the global interconnect. It is assumed in Figure 1.8 that the die edge length may be approximated by the global interconnect length. The greatest leverage in modulating line resistance is in line

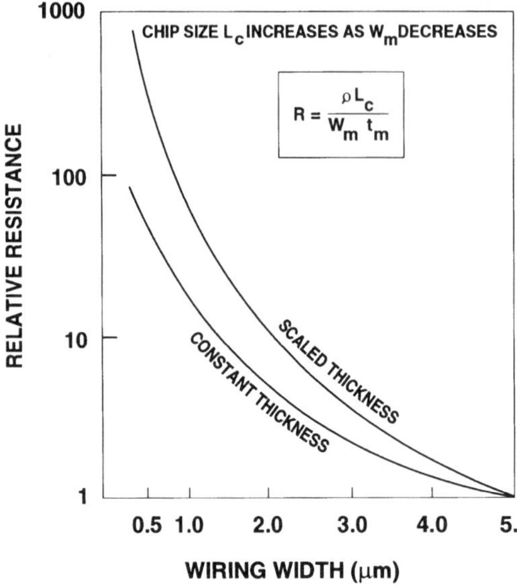


**Figure 1.8** Variation of resistance as a function of interconnect width.[12] (© *1982 IEEE.*)

width and thickness. This really ties in to the earlier discussion of dedicating interconnects to specific applications with specific line widths and thicknesses. Line length also contributes to overall resistance. As chip size increases, the line length increases, resulting in higher resistance. To lower the resistance of long lines, they are segmented into smaller lines and connected together by repeaters or buffers.

**Interconnect capacitance.** Capacitances exist as insulating material sandwiched between conducting films. It is extremely important to know the source and magnitude of these capacitances, as they contribute to RC time delays. Figure 1.9 shows the capacitance environment in a 1-μm technology multilevel interconnect circuit.[4] This figure illustrates capacitive coupling between various interconnects and from interconnects to the substrate. As the packing density increases, the metal lines are placed closer together. Dielectric deposition and planarization become critical in enhancing or reducing capacitances that may exist. As metal pitches decrease, capacitances increase. However, if all cross-sectional dimensions are scaled then the capacitance values remain constant. Processing problems and constraints may also alter capacitances when, for example poor dielectric gap fill and planarization occur. Figure 1.10 presents a schematic view of the major types of capacitances in a dual-level metal interconnect system.

**Interconnect-to-interconnect.** As the metal pitch is reduced [line width ($W_i$), line space ($W_{sp}$)], metal-to-metal capacitance, $C_{ii}$, increases significantly and dominates the overall interconnect capacitance. The para-

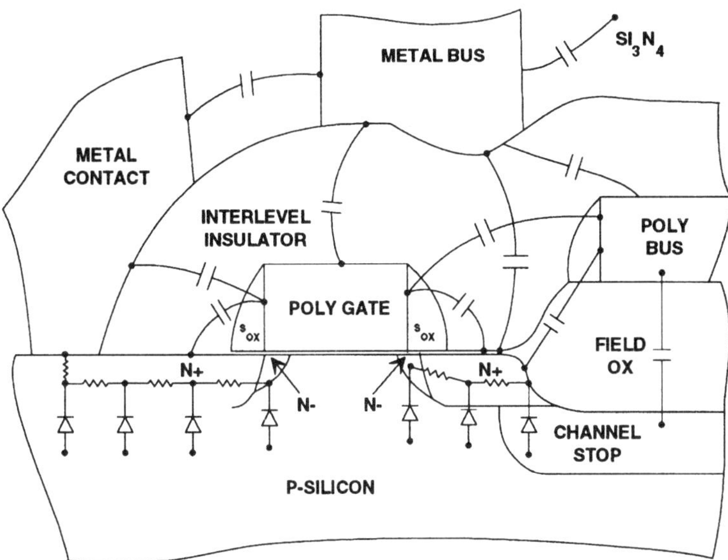


**Figure 1.9** Parasitic capacitance environment in 1-μm MOS technology.[4] *(Courtesy of Van Nostrand Reinhold.)*

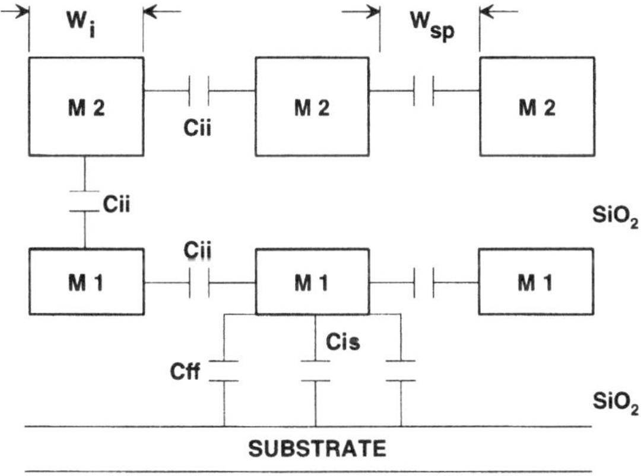


**Figure 1.10** Schematic representation of the components of interconnect capacitances.

sitic capacitance between interconnections leads to degradation in switching speeds and also causes cross talk between adjacent signal lines. To minimize RC time delay and maximize packing density, the interconnect thickness should be held constant as the line width ($W_i$) and line space ($W_{sp}$) are scaled down. The relationship is more complicated.

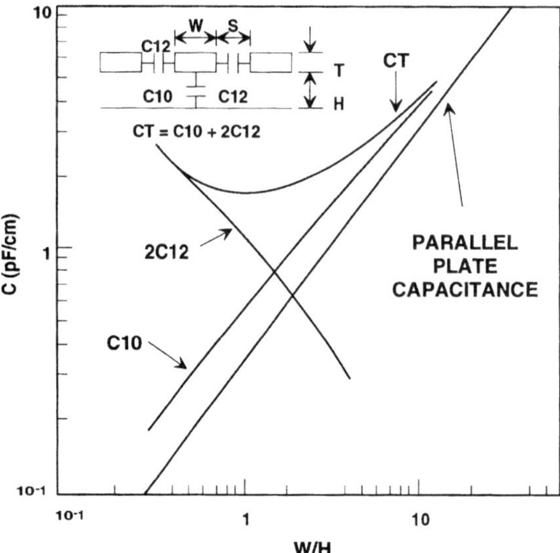


**Figure 1.11** Fringing capacitance variation.[4] (*Courtesy of Van Nostrand Reinhold.*)

**Parallel plate.** Let us focus on the capacitances that are formed from a single interconnect. Consider the capacitances $C_{is}$ and $C_{ff}$, lines originating from M1 in Figure 1.10. Single interconnect capacitance has two components to it: perpendicular and fringing capacitances. The perpendicular capacitance lines, $C_{is}$, are the field lines between the interconnect and the ground plane (substrate). $C_{ii}$ and $C_{ff}$ together are larger than the perpendicular capacitance, $C_{is}$, which is dependent on line width.

**Fringing fields.** The field lines between an interconnect and a ground plane also originate from the interconnect edges. This is in addition to field lines that are perpendicular to the interconnect. This capacitance is independent of the line width. Figure 1.11 illustrates interconnect capacitance variation[4] of a parallel bus line with decreasing pitch. The fringing capacitances are denoted by $C12$ ($C_{ff}$ in this book). It is quite clear that the fringing capacitances contribute significantly to the overall capacitance. The total capacitance reaches its lowest point when the interconnect width and dielectric thickness are equal. However, as this ratio is reduced below approximately 1, the total capacitance starts to increase, primarily due to the fringing capacitances. Dielectric thickness may be increased to decrease the capacitance and therefore the RC time delay. However, such an increase may cause two problems. First, the dielectric thickness approaches the interconnect width, the total capacitance decreases slowly due to fringing fields. Second, the metal width is scaled down, the total interconnect capacitance decreases while the adjacent (metal to metal) capacitance increases. The interconnect capacitance, $C_i$, may be expressed in the following way:

For local interconnects $$C_i \propto K_{ox}\epsilon_0 \frac{L_i W_i}{T_{ox}} \tag{1.8}$$

where $\epsilon_0$ = permittivity in free space
$K_{ox}$ = dielectric constant
$T_o$ = thickness of oxide
$W_i$ = interconnect width
$L_i$ = interconnect length

For global interconnects $$C_i \propto K_{ox}\epsilon_0 \frac{L_g W_i}{T_{ox}} \tag{1.9}$$

where $\epsilon_0$ = permittivity in free space
$K_{ox}$ = dielectric constant
$T_o$ = thickness of oxide
$W_i$ = interconnect width
$L_g$ = average global interconnect length

After scaling the scaled capacitance, $C_i'$, is given by the following:

For local interconnects $$C_i' = S^{-1} C_i \tag{1.10}$$

For global interconnects $$C_i' = S_c C_i \tag{1.11}$$

Let us examine the variation of interconnect capacitance. Figure 1.12*a* and *b* show the variation of total capacitance as a function of metal width with different conductor thickness and dielectric permittivities, respectively.[17] As expected, these plots show the interconnect capacitance decreasing with decreasing metal width and increasing interconnect thickness. Decreasing dielectric permittivities also decreases capacitance.

### 1.3.3 RC time delay cases

From the equations given above for scaled resistance and capacitances, $R_i'C_i'$, delay due to scaling is given by the following formulas:

For local interconnects $$R_i'C_i' = R_iC_i \tag{1.12}$$

For global interconnects $$R_i'C_i' = S^2S_cR_iC_i \tag{1.13}$$

Let us examine the various situations in which the RC time constant may be modulated for local interconnects. The following case study may be expanded for global interconnects also.

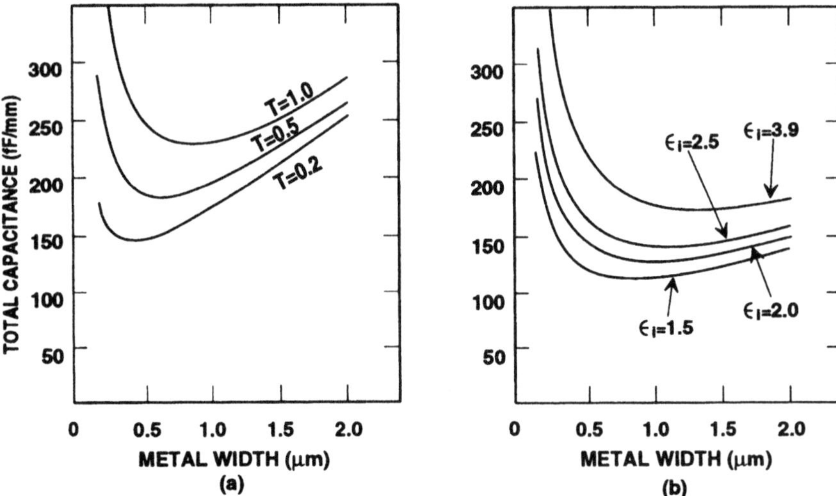


**Figure 1.12** (*a*) Impact of metal width and thickness on interconnect capacitance. (*b*) Impact of metal width and dielectric constant on interconnect capacitance.[17] (© *1990 IEEE.*)

**Case 1.** If all dimensions are scaled equally, then

$$R_i'C_i' = R_iC_i \tag{1.14}$$

This suggests that scaling all the dimensions by the same factor does not change the time delay.

**Case 2.** If interconnect thickness is constant, then

$$R_i'C_i' = S^{-1}R_iC_i \tag{1.15}$$

In this case the scaled time delay is lower than the original time delay.

**Case 3.** If interconnect length is constant, then

$$R_i'C_i' = S^2R_iC_i \tag{1.16}$$

This case suggests that the interconnect length is a dominant parameter in determining the RC time delay of a circuit. If the interconnect length is not scaled, then the RC time delay increases by $S^2$.

**Case 4.** If interconnect width is constant, then

$$R_i'C_i' = R_iC_i \tag{1.17}$$

**Case 5.** If material properties are changed, then

$$R_i'C_i' < R_iC_i \tag{1.18}$$

These five cases present an interesting interconnect scaling situation. If all dimensions are scaled by the same factor then no decrease in RC time delay is realized. These relationships suggest that to minimize RC time delays, an interconnect must be thick, wide, and short. To obtain acceptable RC delays, global interconnects can be made wider and thicker than the first few metal layers. To minimize the resistance increase due to line length increase, the interconnects may be broken up into smaller segments using repeaters, which are explained in Section 1.4.3, below.

### 1.3.4 Interconnect materials

The RC time delay may also be decreased by using conducting materials with low-resistivities and dielectric films with low dielectric constants. The choice and selection of dielectric materials is a little more complicated. It has to be compatible with planarization processes and a host of other requirements. A detailed examination of the desired properties of interconnect and dielectric materials is presented in Chapter 2. The selection of low resistance interconnect materials depends on the application of the interconnect. For example, for decreasing source and drain resistances and polyinterconnect resistance, refractory materials are used. For metal 1 interconnects, aluminum (Al) and its alloys are widely used, but other materials like tungsten (W) are also used. For higher level interconnects, Al and its alloys are the main interconnect materials. To make the interconnects more robust, Al films are used along with more EM-resistant materials like titanium (Ti). Table 1.4 compares various interconnect materials,[15] specifically for lower levels such as junction and local interconnects.

### 1.3.5 Scaling rules

The ideal scaling rules[6,7] for both local and global interconnections are summarized in Tables 1.5 and 1.6.

### 1.3.6 Modified scaling

In previous sections, the ideal scaling concepts were discussed. On closer examination of the five RC time delay cases, it appears that deviating from ideal scaling, and scaling some parameters differently than others may be advantageous. In case of ideal scaling, all linear and vertical dimensions were scaled by $1/S$. Let us examine other scaling techniques[6] that change the RC delay and packing density. Packing density (PD) refers to the number of transistors that can be formed in a given area. That is,

**TABLE 1.4 Properties of Interconnection Materials**[15]

| Materials | Bulk resistivity | Thick film resistivity (μohm/cm) | Melting point (°C) | Chemical compatibility | Thermal oxidation |
|---|---|---|---|---|---|
| Poly-Si | 250 | 500 | 1410 | Yes | Yes |
| Al | 2.26 | 3 | 660 | ? | No |
| W | 5.5 | 10 | 3410 | ? | No |
| Mo | 5.0 | 10 | 2610 | ? | No |
| $WSi_2$ | 12.5 | 30 | 2165 | Yes | Yes |
| $MoSi_2$ | 21.6 | 40 | 2050 | Yes | Yes |
| $TaSi_2$ | 8.5 | 35 | 2170 | Yes | Yes |
| $TiSi_2$ | 13 | 15 | 1540 | ?* | Yes |

*$TiSi_2$ is etched severely in HF.
SOURCE: © 1991 IEEE.

**TABLE 1.5 Ideal Scaling of Local Interconnects**

| Interconnect parameter | | Ideal scaling |
|---|---|---|
| Dimensions [width ($W_i$), length ($L_i$), spacing ($W_{sp}$), thickness ($T_i$), dielectric thickness ($T_{ox}$)] | | $1/S$ |
| Line resistance | $R_i = \dfrac{\rho L_i}{W_i T_i}$ | $S$ |
| Capacitance: interconnect (i) to substrate (s) | $C_{is} = \dfrac{K_{ox}\epsilon_0 W_i}{T_{ox}}$ | $1/S$ |
| Capacitance: between lines | $C_{ii} = \dfrac{K_{ox}\epsilon_0 T_i L_i}{W_{sp}}$ | $1/S$ |
| RC delay ($T$) | | 1 |
| Voltage drop ($IR$) | | 1 |
| Current density ($J$) | | 1 |

$$PD \propto 1/A$$

$$\propto 1/WL \tag{1.19}$$

For ideal scaling

$$PD \propto S^2 \tag{1.20}$$

where $PD$ = packing density
$W, L$ = lateral dimensions (width, length)
$S$ = scaling factor

Both RC time delay and packing density have to be optimized during circuit design. The number of arrows indicates the relative magni-

**TABLE 1.6 Ideal Scaling of Global Interconnects**

| Interconnect parameter | | Ideal scaling |
|---|---|---|
| Dimensions [width ($W_i$), length ($L_i$), spacing ($W_{sp}$), thickness ($T_i$), dielectric thickness ($T_{ox}$)] | | $1/S$ |
| Global line length, $L_g$ | | $S_c$ |
| Line resistance | $R_i = \dfrac{\rho L_i}{W_i T_i}$ | $S^2 S_c$ |
| Capacitance: interconnect (i) to substrate (s) | $C_{is} = \dfrac{K_{ox}\epsilon_0 L_g W_i}{T_{ox}}$ | $S_c$ |
| Capacitance: between lines | $C_{ii} = \dfrac{K_{ox}\epsilon_0 T_i L_g}{W_{sp}}$ | $S_c$ |
| RC delay ($T$) | | $S^2 S_c^2$ |
| Voltage drop ($IR$) | | $SS_c$ |
| Current density ($J$) | | $S$ |

tude of scaling. The following discussion is applicable for local interconnects. Modified scaling schemes for global interconnects are more complicated due to the inclusion of chip size in the RC calculations. Using the equations for capacitance and resistance, the RC time delay for each of the three schemes can be calculated.

**Scheme 1 (ideal).** If the lateral/vertical dimensions are scaled equally by $1/S$, then

$$\text{RC scaling factor is } 1$$

**Scheme 2 (quasi-ideal).** If the lateral dimensions are decreased more than the vertical dimensions, where the interconnect length/width/space $= 1/S$, the thickness of the interconnect $= 1/\sqrt{S}$, and the thickness of the dielectric $= 1/\sqrt{S}$, then

$$C' \propto 1/S$$

$$R' \propto \sqrt{S}$$

Therefore

$$R'C' \propto \sqrt{S}/S \propto 1/\sqrt{S} \tag{1.21}$$

**Scheme 3 (constant *R*).** If the lateral/vertical dimensions are scaled equally by $1/\sqrt{S}$ except interconnect length which is scaled by approximately $1/\sqrt{S}$, then

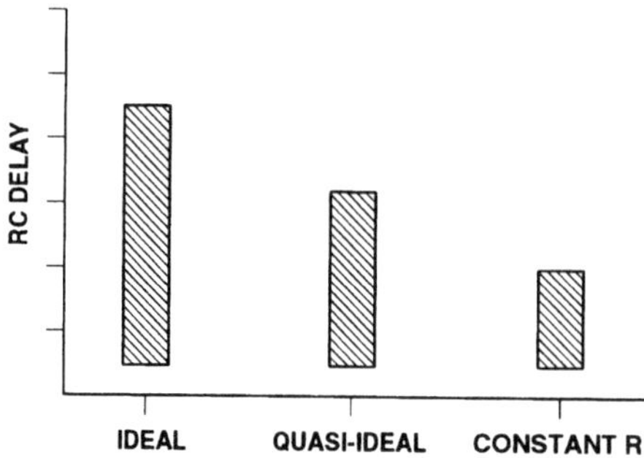


**Figure 1.13** Reduction in RC delay with modified scaling schemes.

$$C' \approx 1/S$$

$$R' : \text{not scaled (constant)}$$

Therefore

$$R'C' \approx 1/S \tag{1.22}$$

Let us compare these scaling schemes with each other to illustrate how RC time delay varies. Figure 1.13 shows that by varying the scaling scheme, it is possible to reduce RC delays in interconnects.

### 1.3.7 Reliability scaling[18]

Let us examine here how scaling would impact metal electromigration performance. Electromigration is the movement of ions of the conducting material itself when current is allowed to pass through it. Electromigration failure is a wearout mechanism and has a time-dependent nature. To determine the quality of metal interconnects, devices are subjected to certain conditions that will accelerate EM failures. The objective of performing accelerated testing is to extrapolate the failure rate for a given set of metal interconnects over time and operating conditions. Scaling device dimensions imposes more stringent reliability conditions on the devices.

Electromigration, that is, mean time to fail (MTTF) for interconnects and contacts may be expressed as

$$\text{MTTF} \propto J^{-2} e^{E/KT} \tag{1.23}$$

where $J$ = electric current density
$E$ = activation energy
$K$ = Boltzman's constant
$T$ = absolute temperature

**TABLE 1.7 Scaling Rules for Electrical Parameters[18]**

| Electrical parameter | Scaling factor |
|---|---|
| Supply voltage | 1 |
| CMOS RMS current : contact/line | $S$ |
| Metal current density | $S^3$ |
| Contact area current density | $S^3$ |

**TABLE 1.8 Scaling Rules for Electromigration[18]**

| Electromigration scaling ratio ($MTTF_s/MTTF_u$) | Scaling factor |
|---|---|
| Metal line | $1/S^7$ |
| Contact (open) | $1/S^9$ |
| Contact (junction leakage) | $1/S^7$ |

The current density is the cross-sectional area of the conductor and the current flowing through it. Under ideal scaling conditions with voltage being constant, the current increases linearly by the scaling factor $S$. The current density scales by $S^3$ when the exponent is 2 for Al. These are shown in Table 1.7. Applying the scaling rules for electrical parameters in the EM MTTF relationship, the ratio of the scaled to the unscaled MTTF values may be expressed as shown in Table 1.8.

The scaling rules from these tables suggest that as transistor and interconnect dimensions are scaled down, the wearout failure mechanism (such as EM) rapidly increases. The acceleration of the wearout failures is worse when the dimensions are scaled down and the operating voltages are not. To overcome the EM handicap of Al conductors, layered interconnects are used to enhance the EM performance.

## 1.4 RC Time Delay

It is critical to model RC delays in circuits[19–27] to accurately predict the effects of scaling and interconnect layout/routing. As the transistors and interconnects are laid out on a chip, the resistances and capacitances may be extracted and visualized in the form of a network of resistances and capacitances. The objective is to determine the overall RC delay of the circuit. RC delays are cumulative in nature; therefore, the effects of resistance and capacitance accumulation in networks and trees must be considered during RC modeling. Once the total RC delay for a circuit is determined, techniques must be developed to reduce it if the speed performance targets are not met for the IC. There are few methods of reducing RC delays in circuits. The selection of appropriate interconnect and dielectric material is one option that was discussed in Section 1.3. A technique discussed here is to partition the interconnects into smaller segments to reduce the effective RC delay. The analysis of RC modeling and RC delay reduction techniques is used in determining an efficient interconnect layout on a chip.

### 1.4.1 General RC networks

What is RC time delay? Electrically it is the product of resistance and capacitance. This means the delay time may be defined as the time taken for the signal at the load to reach 90 percent of the input signal.[10] The threshold for logic devices is approximately 50 percent. The concepts associated with RC delays are discussed in the following step-by-step approach.

**Step 1: RC circuit elements.** Let us first look at the whole multilevel interconnect circuit from a block diagram perspective. A typical circuit consists of the following functional blocks: driver, interconnect, and load, as shown in Figure 1.14. In Figure 1.14, the main circuit elements for each block are highlighted. The driver block's main contribution is resistance. Both resistance and capacitance are major elements for an interconnect connecting the driver transistor to the load. The primary circuit element of interest from a load is its capacitance. How do all these interact? When turned on, a driver transistor drives current through an interconnect to a load transistor. The driver transistor contributes to the delay due to its own resistance, $R_{DR}$, and the signal propagates through the interconnect, $(R_I C_I)$. The terminal point is the load capacitor, $C_L$. From this block diagram and from previous discussions, it is intuitively clear that when the interconnect resistance and/or capacitance increase, RC delay increases. There is a loss in signal strength due to interconnect parasitic capacitances. Obviously, in

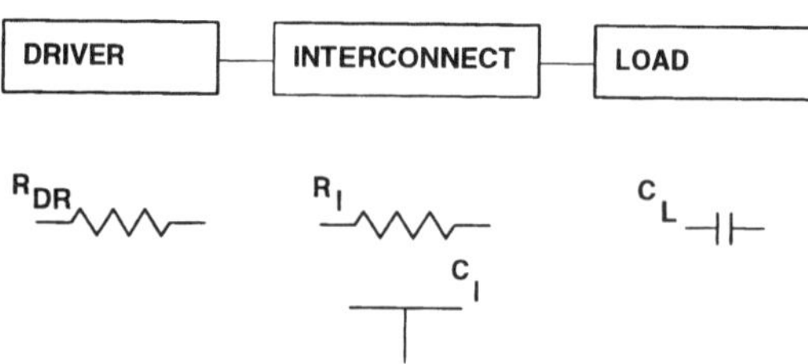


**Figure 1.14** A schematic representation of multilevel interconnect functional blocks and equivalent circuit elements.

circuits that have been designed for high switching speeds, the RC delay is a major component for speed degradation. When an interconnect has a very high RC time constant per unit length, it will be not be possible to obtain high clock frequencies by the clock driver.

**Step 2: RC modeling circuits.**[6,28] In MOS circuits, calculation of signal delay is quite complex as a given invertor or logic node may drive several gates. In some cases, long interconnects are used whose distributed resistance and capacitance may not be negligible and contribute to the signal delay. Therefore, the signal delay may be modeled using two types of circuits, lumped and distributed RC circuits, as shown in Figure 1.15. The objective in each of these circuits is to determine the output voltage, which is a function of RC, as the capacitor starts to discharge.

In lumped circuits, all the capacitances can be lumped together and the all the resistances except the pullup/load transistor resistance, $R_{dr}$, can be neglected. The circuit response is easy to calculate for a lumped circuit and is given by

$$V_{OUT}(t) = 1 - e^{-t/RC} \tag{1.24}$$

where $V_{OUT}(t)$ = the output voltage of the circuit
$R$ = resistance of pullup/load transistor
$C$ = the total capacitance
$t$ = time

From this equation it is possible to determine the time when $V_{OUT}(t)$ reaches some specified critical voltage. This equation gives exponen-

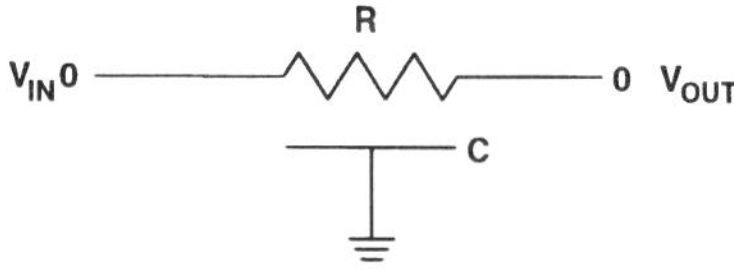


(A) DISTRIBUTED

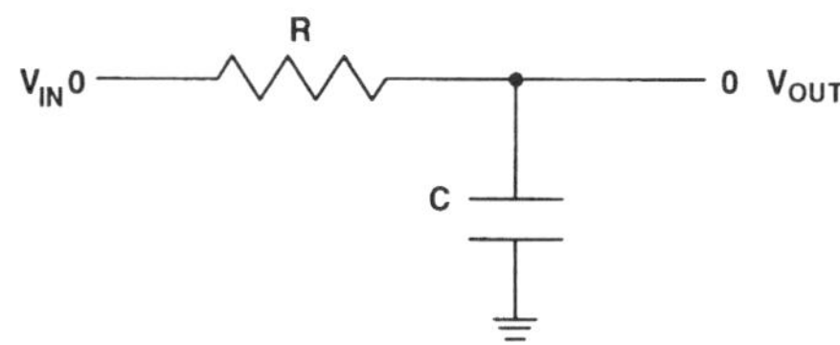


(B) LUMPED

**Figure 1.15** Two basic RC modeling circuits.[6] (*Reprinted with permission of Addison-Wesley Publishing Co., Inc.*)

tial decay in signal strength, which is dependent on the capacitance of an interconnect. Lumped circuits may be arranged in several ways.[29] For example, one approximation may be like the lumped circuit shown in Figure 1.15*b*. Another representation may be in the form of a T-section, as shown in Figure 1.16.

When the interconnect resistances are comparable to that of pullup/load transistor, they cannot be neglected. In such a case the circuit response is easy to calculate. In order to determine the time delay, distributed RC networks may be approximated by lumped circuits.

**Step 3: RC network.** The approximation of distributed circuits network is achieved by combining several lumped circuits together as shown in Figure 1.17 to form a RC network. The time delay[28] in a RC circuit may be expressed as

$$T_d = \tfrac{1}{2} RCI^2 + RC_L I + R_D CI + R_D C_L \qquad (1.25)$$

where $R/C$ = interconnect parasitics/unit length
$R_D$ = driver resistance
$C_L$ = load capacitance
$I$ = interconnect length

From this equation the delay time may be determined as a result of *R, C, I* scaling. Expanding on this equation, analysis of the distrib-

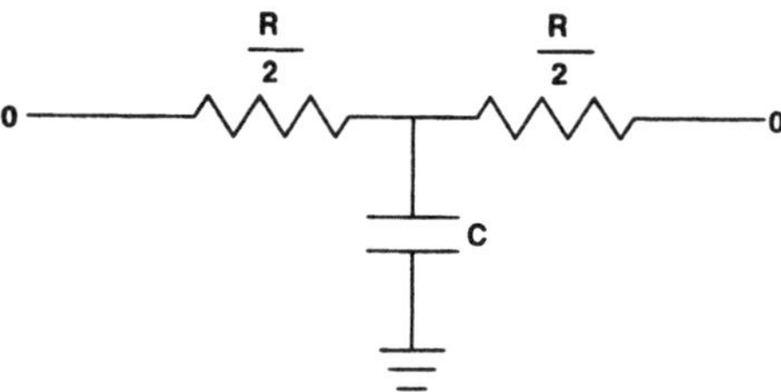


**Figure 1.16** A T-shaped lumped RC circuit approximation of a distributed RC circuit.[29] (*© 1985 IEEE.*)

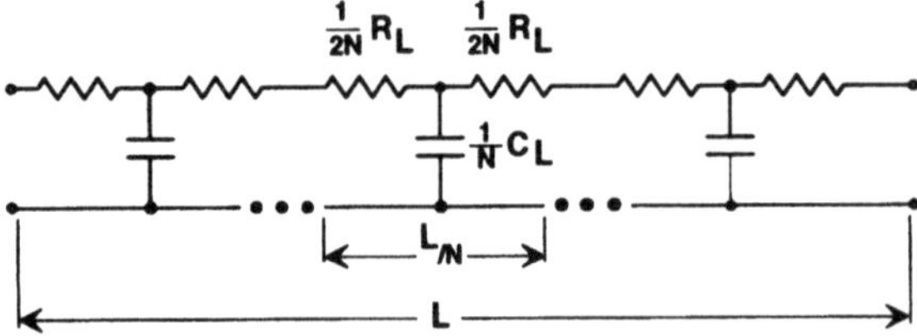


**Figure 1.17** A RC network approximated by a cascade of T-sections.[28] (*Courtesy of John Wiley & Sons, Inc.*)

uted RC network under a step voltage excitation can be determined. The time delays calculated for output voltages to rise from 0 to 50 and 90 percent and the equations are as follows.[6] For 0 to 90 percent rise

$$T_{90\%} \approx (2.3R_D + R_i)C_i \tag{1.26}$$

where $R_D$ = resistance of the driver transistor
$R_i$ = resistance of the interconnect
$C_i$ = capacitance of the interconnect

This holds true when the load capacitance is much less than the interconnect capacitance. Similarly, for 0 to 50 percent rise

$$T_{50\%} \approx (0.7R_D + 0.4R_i)C_i \tag{1.27}$$

**Step 4: RC trees.** Imagine a CMOS IC with an array of transistors, some drivers, and others loads, interconnected to each other. From such an array we can extract resistances and capacitances forming an RC tree with branches, as shown by the block diagram in Figure 1.18. Each of these blocks may contain lumped and distributed RC circuits; for an effective circuit design and modeling we need to determine the RC delay in an RC tree. The total time delay is a sum of its individual components. That is, time delays from A to D and C may be expressed as

$$T_{AD} = T_{AB} + T_{BD} \tag{1.28}$$

$$T_{AC} = T_{AB} + T_{BC} \tag{1.29}$$

### 1.4.2 Impact on RC delay

As shown above, interconnect resistance and parasitic capacitances modulate RC delays in ICs. In addition, line width and length impact RC delay by virtue of their effect on interconnect resistance and capacitance. Figure 1.19 shows the delay time for various line widths and line lengths with a 100-ohm driver.[3] As the length increases, the RC time delay dominates. For local interconnects the relationship between delay and interconnect length is nearly linear with total $R$ approximately equal to $R_{dr}$ and total $C$ equal to $C_i$. In such cases, the main factors that dominate RC delay are the interconnect capaci-

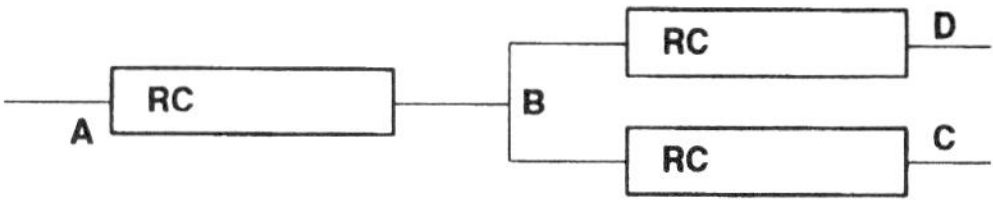


**Figure 1.18** A schematic block representation of RC tree network.

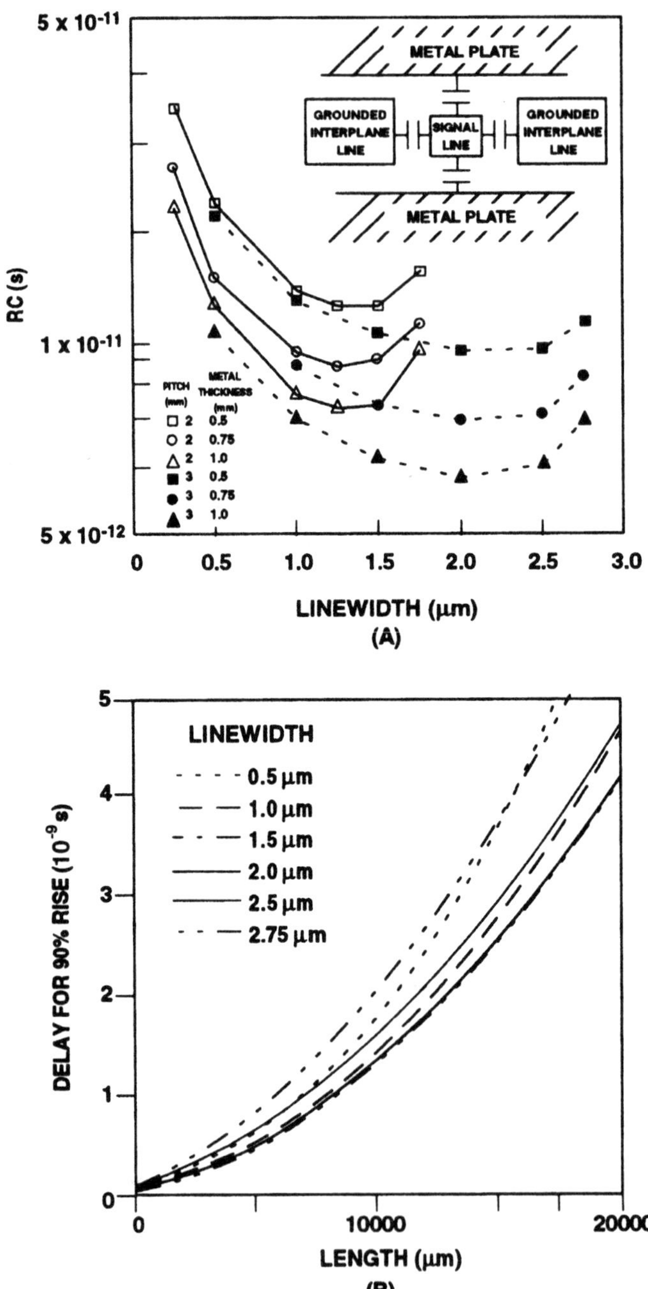


**Figure 1.19** RC delay as a function of (*a*) interconnect length[3] and (*b*) interconnect width.[3] (*Courtesy of Solid State Technology.*)

tance and the driver's output resistance. Let us elaborate on this relationship since the RC and interconnect width relationship is quite interesting. When the width is scaled down, a critical value of interconnect width is reached when the RC delay increases drastically. When the interconnect width is less than the critical value, the dominating factors that modulate delay are capacitance and resistance of the interconnect. Although dependent on the scaling scheme used, fringing capacitance effects impact the rate of change of capacitance and resistance. Therefore, the RC delay per unit length increases. When the interconnect width is greater than the critical width, the driver's output resistance and interconnect capacitance modulate the delay.

### 1.4.3 RC time delay reduction

From earlier discussion it is evident that one way of reducing RC time delay in interconnects is to decrease their length. Other methods are using low-resistivity materials and using a hierarchy of interconnect levels with thicker and wider lines in the upper levels. Decreasing the length of the interconnect may not be possible, especially when the die size increases. But the RC time delay may be reduced by using buffers or repeaters that provide the same effect as interconnect length reduction. Table 1.9 summarizes the effect on RC time delay for an interconnect of length $L$ and compares that to the same interconnect that has been divided in $k$ equal parts of length $L/k$. The cumulative RC delay for the partitioned interconnect is reduced as compared to delay of the original interconnect.

The technique of reducing the delay is quite effective for long global interconnects that may be susceptible to large RC delays. Partitioning the interconnect into smaller segments and reducing the delay is possible as long as the repeaters/buffers do not contribute significantly to the delay. For total delay calculations, the delay due to the repeaters and buffers must also be taken into consideration. Moreover, the repeaters/buffers must be designed to occupy the minimum possible

**TABLE 1.9 Effect on RC Delay**

| Parameter | Full length | Divided length ($k$ equal parts) |
|---|---|---|
| Length | $L$ | $L/k$ |
| Resistance | $R$ | $R/k$ |
| Capacitance | $C$ | $C/k$ |
| RC time delay | $RC$ | $RC/k^2$ |
| Total RC time delay | $RC$ | $R'C' = (RC/k^2) * k$ $= RC/k$ |

**TABLE 1.10 Summary of RC Reduction Techniques**

| RC reduction technique | Propagation delay[6] | Remarks |
|---|---|---|
| Minimum-size repeater | $T_{50\%} = 0.7R_oC_i + 1.1\sqrt{R_oC_oR_iC_i} + 0.7R_a$ | 1. Divides interconnect length into equal segments<br>2. RC delay varies linearly with length<br>3. Effective when interconnect RC is large |
| Optimal-size repeaters | $T_{50\%} = 2.5\sqrt{R_oC_oR_iC_i}$ | 1. Driver capability enhanced, reduces output resistance<br>2. Current drive capability proportional to device $W/L$<br>3. Effective when interconnect RC is large |
| Cascade input drivers | $T_{50\%} = 0.7eR_oC_o\ln\left(\frac{C_i + C_L}{C_o}\right) + 0.4R_{iC_i}$ | 1. More effective than previous methods when interconnect RC is not very large<br>2. Used to drive large capacitive loads<br>3. Size of drivers increases gradually |

die area. A summary of the different RC reduction techniques is given in Table 1.10.

## 1.5 Interconnect Layout

In MOS technology, three types of interconnects may be used: diffusion, polysilicon, and metal. Long diffusion lines are not used in circuit design due to large capacitance per unit length. This is detrimental to circuit performance. Therefore, interconnect routing is limited to polysilicon and multiple metal layers. The common technique in interconnect routing has been to ensure that no two interconnects of the same layer cross each other. Therefore, each layer is assigned a particular direction, either horizontal or vertical, so that the only possible connection between the two is through a via. This horizontal and vertical arrangement is very suitable for multilevel interconnects, but the requirement for a large number of vias (layer-to-layer connections) is a problem with this arrangement. Via minimization in interconnect layout needs to be addressed to obtain efficient and effective layout designs. Minimizing the number of vias will help increase the overall yield as there will be less opportunity for via-related defects.

Using multiple level of interconnects and optimizing interconnect routing has a significant impact on the performance of ICs. In 1991, Intel developed[30] a CMOS 100-MHz microprocessor that was architecturally identical to the Intel486™. The success in obtaining such high frequency was attributed to an aggressive reduction of interconnect length to reduce area and parasitics and an efficient interconnect routing scheme. Of course, improved process technology tailored for three levels of metal also contributed to the development of this microprocessor. Figure 1.20 and Table 1.11 compare the circuit performance of 1- and 0.8-μm 2/3 metal technologies. The increase in parasitic capacitances was offset by area reduction and an efficient routing system. Table 1.11 highlights various output parameters that contributed to the development of a 100-MHz microprocessor.

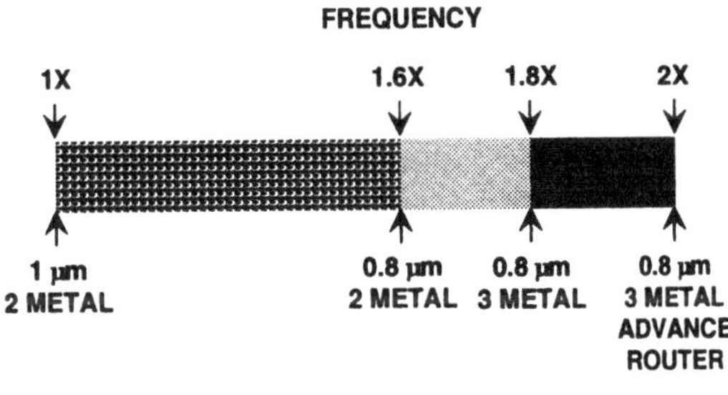


**Figure 1.20** Impact of interconnect routing on device performance.[30] (© *1991 IEEE.*)

**TABLE 1.11 Output Parameters for High-Frequency Microprocessor[30]**

| Auto router random logic | 1 μm, 2 metal | 0.8 μm, 3 metal | Ratio |
|---|---|---|---|
| $C_{tot}$ (pf) | 274 | 239 | 0.87 |
| $C_{int}$ (pf) | 122 | 64 | 0.52 |
| Z total (μm) | 61,888 | 51,491 | 0.83 |
| Native delay (ns) | 10.51 | 5.90 | 0.56 |
| Tuned delay (ns) | 10.51 | 5.25 | 0.50 |
| Area (mm$^2$) | 1.91 | 0.89 | 0.42 |

SOURCE: © 1991 IEEE.

### 1.5.1 Design integration[31]

Design of advanced devices is a complex task. A highly productive technique is design integration. This involves using predesigned functions in one large integrated circuit. A set of predefined functions can be put in a block (also referred to as megacells, compilers, or macros) and several blocks may be integrated on a large chip to obtain the desired circuit functionality. Once the required number of blocks are defined, the next step is designing interface logic that makes these blocks work together. The last step is defining an efficient routing scheme between various blocks with a minimum number of vias. The use of functional blocks to build an IC is widely used for application-specific integrated circuits (ASIC). Using predesigned components, such as blocks helps

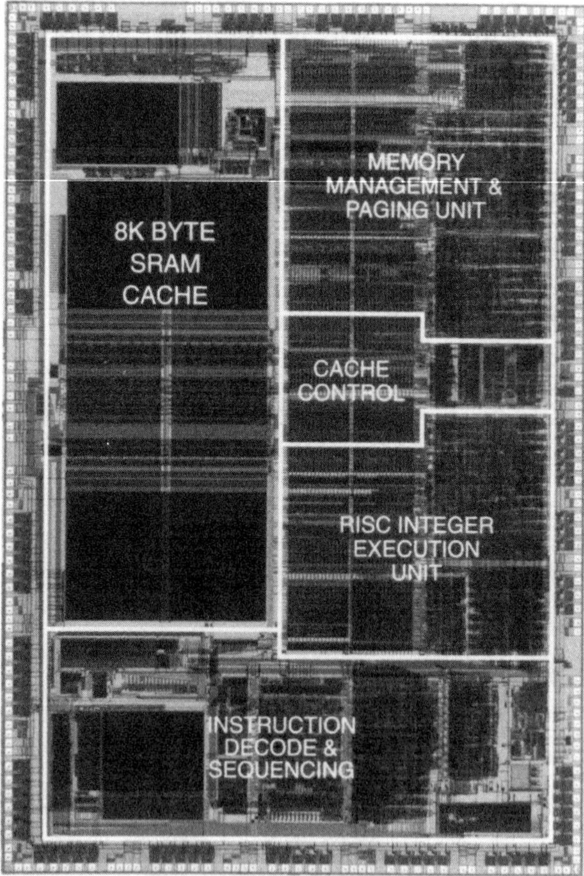


**Figure 1.21** Intel486™ chip showing various functional blocks.

reduce the design effort needed for new-generation devices. Figure 1.21 shows an Intel486™ chip with various blocks highlighted.

Most automated design tools that are used for interconnection layout perform the following functions:

1. Fit all blocks required to achieve circuit functionality in a given die area
2. Optimize shape and relative position of each block in a given circuit
3. Determine and optimize intrablock interconnect layout
4. Determine and optimize interblock interconnect layout

Figure 1.22 is a generic illustration of blocks in the formation of an IC. Each block may have a separate function. For example, in one type of device, a block may be dedicated to SRAM circuitry and in other types more than one block may be dedicated to memory and the rest for other functions. By incorporating functional blocks in various designs, the designers have the opportunity to thoroughly comprehend the robustness of the block design. Once the blocks have been placed, the space between them is designated as a channel. Channels are used for interconnect routing. As explained earlier in this chapter, interconnect routing may be divided into two parts—local and global—depending on the interconnect function. Global routing is used for interblock connections and so the lengths are long. As a result, the final routing area is significantly dependent on global routing paths.

### 1.5.2 Layout verification

Layout verification is a very difficult and cumbersome process when done manually. New automatic verification tools are offering an attractive verification process. Figure 1.23 shows an automatic layout verification system that has the capability to perform various tasks and provide the necessary information to the designer, who can then release the design for production. The main features of this system are numbered on the figure. The verification process uses data from circuit- and technology-related data bases. Design rules are verified (1) for any violations. For example, metal interconnect overlap of contacts may be specified as part of the design rules. Design violations with regard to overlap may cause processing problems, so it important to understand how much overlap is detrimental to the product and if the overlap specified is compatible with the variation in reticle exposure and layer misalignment. Electrical design rules (2) are also verified. These pertain to layer assignment and interconnect crossing, missing vias, etc. Having checked the physical and electrical design

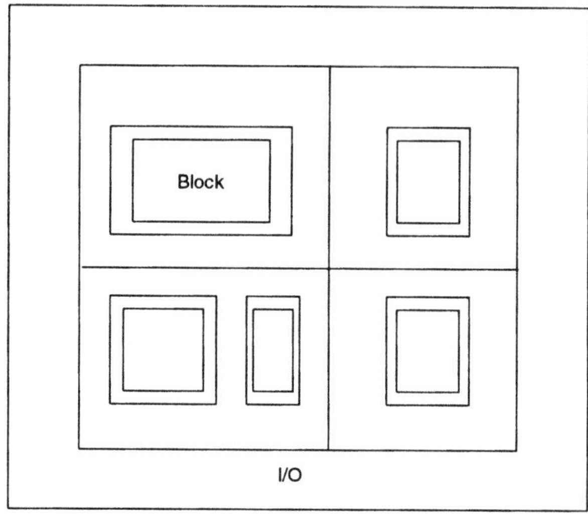

Figure 1.22 Schematic illustration of an IC block layout.

rules, the next step in layout verification is in connectivity (3). Are the interconnects connected to the proper pins? The actual connectivity pattern has to be compared with connection lists to ensure the desired circuit elements are connected. Other programs (4) check various device components and then extract various parameters (5), such as parasitic capacitances. The design effort involved in interconnect layout and its verification may be reduced depending on the nature of the device being designed. The use of standard cells, building blocks, etc. are steps in this direction.

## 1.6 Summary

This chapter has provided an overview of the main design concerns associated with designing circuits with multilevel interconnect technology. The discussion on scaling and RC delay in interconnections has set the stage for the discussion on the process, manufacturing, and reliability aspects of multilevel interconnect technology. The challenging and complex design tasks originate in response to a need for satisfying or creating a market demand. This includes packing more features in every new generation chip with the use of new device and submicron process technologies. Starting from product development

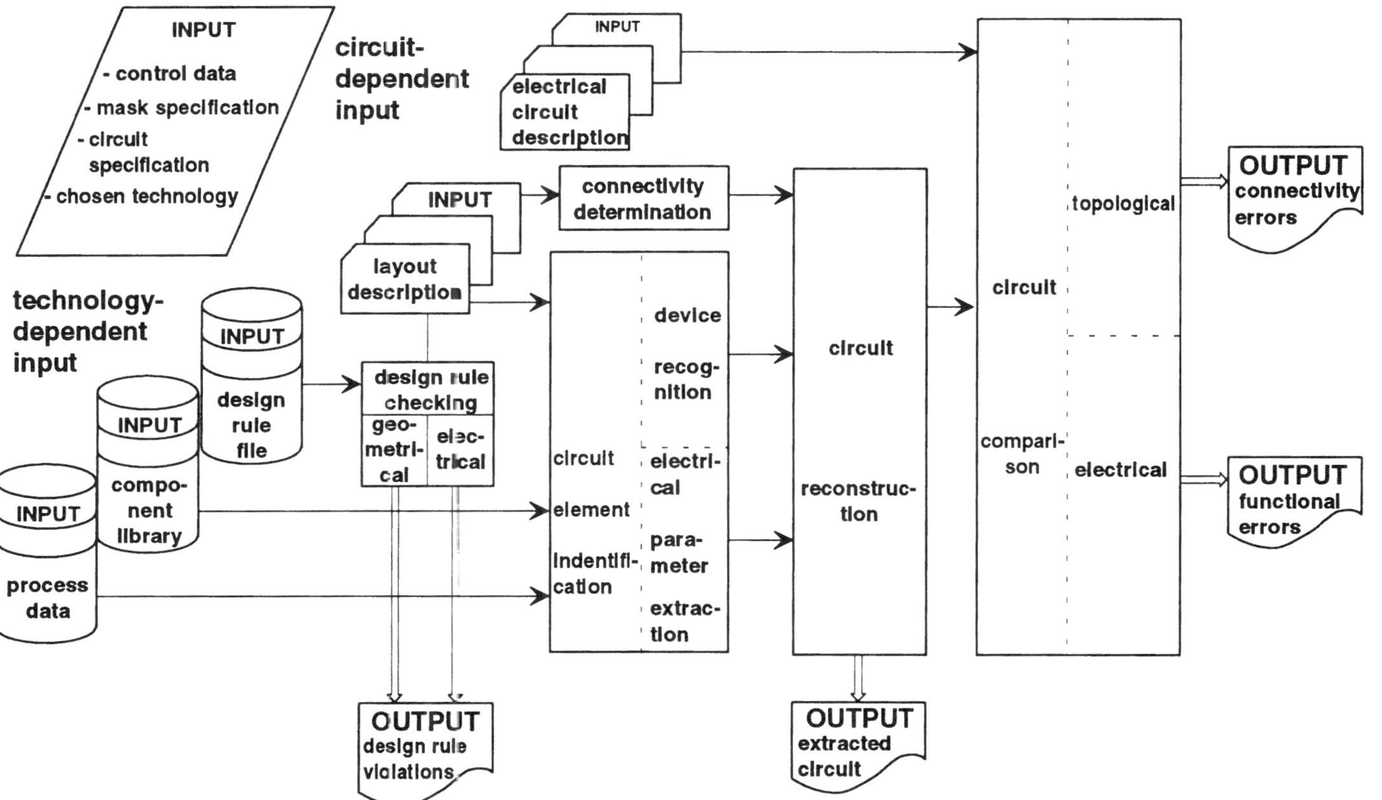


**Figure 1.23** Integrated circuit layout verification system.[31] (*Courtesy of Van Nostrand Reinhold.*)

issues, the focus shifted to mastering the four areas in design methodology: functionality, testability, manufacturability, and marketability. New device technologies, like BiCMOS, are being used, to enhance the performance of devices. In addition to this, the performance of devices is also modulated by interconnects, specifically the RC delays associated with them. Reduction of RC is one of the primary goals that has implications on process development (new materials, dielectric deposition, and planarization). As chips get larger or when the designs go through a compaction exercise and more functional blocks are placed on them, optimizing interconnect routing becomes a critical part of the design process.

## References


1. W. J. McClean (ed.), *Status 1992: A Report on the Integrated Circuits Industry,* Integrated Circuit Engineering Corp., 1992.
2. R. A. Haken et al., "Solving the process integration challenges of BiCMOS," *Semiconductor International,* May 1989.
3. S. R. Wilson et al., "A four metal layer, high performance interconnect system for bipolar and BICMOS circuits," *Solid State Technology,* November 1991.
4. P. Chatterjee, "CMOS devices," in G. Rabbat (ed.), *Handbook of Advanced Semiconductor Technology and Computer Systems,* Van Nostrand Reinhold Co., 1988.
5. Y. A. El-Mansy and W. M. Siu, "MOS technology advances," in G. Rabbat (ed.), *Handbook of Advanced Semiconductor Technology and Computer Systems,* Van Nostrand Reinhold Co., 1988.
6. H. B. Bakoglu, *Circuits, Interconnections and Packaging for VLSI,* Addison-Wesley Publishing Co., 1990.
7. R. H. Dennard et al., "Design of ion-implanted MOSFET's with very small physical dimensions," *IEEE J. Solid State Circuits,* Vol. SC-9, 256, 1974.
8. R. R. Troutman, *IEEE Trans. Electron Devices,* Vol. ED-26, 461, 1979.
9. G. Declerck et al., "MOS technology for VLSI," *Microelectron. Reliab.,* Vol. 24, No. 2, 1984.
10. D. S. Gardner et al., "Interconnection and electromigration scaling theory," *IEEE Trans. Electron Dev.,* Vol. ED-34, No. 3, March 1987.
11. Y. Pauleau, "Interconnect materials for VLSI circuits," *Solid State Technology,* February 1987.
12. D. J. McGreivy, "Interconnections/gates in VLSI technologies," in *VLSI Technologies through the 80s and Beyond,* IEEE Computer Society Press, 1982.
13. D. Pramanik and G. Spadini, "Interconnections in application specific VLSI," *European Transactions on Telecommunications and Related Technologies,* Vol. 1, No. 2, March/April 1990.
14. P. S. Ho, "VLSI interconnect metallization," *Semiconductor International,* August 1985.
15. K. Saraswat, "Effect of scaling of interconnections on the time delay of VLSI circuits," *IEEE Trans. Elect. Dev.,* Vol. ED-29, No. 4, April 1982.
16. M. B. Small and D. J. Pearson, "On-chip wiring for VLSI: status and directions," *IBM J. Res. Develop.,* Vol. 34, No. 6, November 1990.
17. Y. Ushiku et al., "Design guidelines for deep-sub-micrometer interconnections," *VMIC,* 1990.
18. M. H. Woods and B. L. Euzent, "Reliability in MOS integrated circuits," *IEDM,* 1984.
19. T. Lin and C. A. Mead, "Signal delay in general RC networks," *IEEE Trans. CAD,* Vol. CAD-3, No. 4, October 1984.
20. J. Rubinstein et al., "Signal delay in RC tree networks," *IEEE Trans. CAD,* Vol. CAD-2, No. 3, July 1983.

21. S. Su et al., "A simple and accurate node reduction technique for interconnect modeling in circuit extraction," *IEEE Intl. Conf. Computer Aided Design: ICCAD-86,* 1986.
22. C. J. Scott, "Transmission characteristics of narrow line-width interconnections on silicon substrates," *VMIC,* 1989.
23. D. Zhou et al., "Interconnection delay in very high speed VLSI," *IEEE Trans. Circuit Systems,* Vol. 38, No. 7, July 1991.
24. M. Jackson and E. Kuh, "Estimating and optimizing RC interconnect delay during physical design," *IEEE Intl. Symp. on Circuits and Systems,* 1990.
25. C. Wu and M. Shaiu, "Accurate speed improvement techniques for RC line and tree interconnections in CMOS VLSI," *IEEE Intl. Symp. on Circuits and Systems,* 1990.
26. I. Koren et al., "Designing interconnection buses in VLSI and WSI for maximum yield and minimum delay," *IEEE Journal of Solid State Circuits,* Vol. 23, No. 3, June 1988.
27. C. Svensson and M. Afghahi, "On RC line delays and scaling in VLSI systems," *Electronics Letters,* Vol. 24, No. 9, April 1988.
28. J. R. Brews, "Electrical modelling of interconnections," in R. K. Watts (ed.), *Submicron Integrated Circuits,* John Wiley & Sons, 1989.
29. T. Sakurai, "Approximation of wiring delay in MOSFET LSI," *IEEE Journal of Solid State Circuits,* Vol. SC-20, No. 6, December 1985.
30. J. Schutz, "A CMOS 100 MHz microprocessor," *IEEE Intl. Solid State Circuits Conf.,* 1991.
31. R. M. J. M. Otten, "VLSI layout," in G. Rabbat (ed.), *Hardware and Software Concepts in VLSI,* Van Nostrand Reinhold Co., 1983.

# Chapter 2

# Process Development

## 2.1 Introduction

Multiple levels of metallization are needed when, for technical and economic reasons, interconnects cannot be laid out efficiently in a single level and therefore need to be stacked one on top of the other.[1–19,125] The technical issues in interconnect layout stem from increased density of transistors that need to be connected, while economic issues pertain to reducing silicon (Si) real estate by stacking the metal interconnects.

This chapter examines the fabrication processes required for transferring product designs to Si. The main objective of this chapter is to present a simplified and logical flow for multilevel interconnect processing. Fabrication process technology must keep pace with constant advances in product designs for successful high-volume manufacture of new products. Such a developmental pace in process technology requires evaluation of existing process methodologies and development of new ones, often using a process technology driver such as a dynamic random-access memory (DRAM) device.

In the field of process technology development, new lithography processes are needed that are capable of extending current optical lithography limits and pushing them beyond the optical spectrum. Etching fine patterns in different substrates with severe restrictions on critical dimensions and etch bias are some of the requirements for plasma etching. On the dielectric deposition front, new types of insulators or combinations of insulators need to be further improved to provide conformal coverage of the device structures and fill ever-decreasing gaps between the metal interconnects. Dielectric materials are closely coupled with the planarization requirements and techniques. A complete global planarization is required due to lithography and metal step coverage constraints as the device geometries migrate to the levels of half-micron and below. Metallization requirements have become very significant as the interconnects occupy a significant portion of the Si real estate and have a strong influence on overall reliability of the device.

## 2.2 Multilevel Interconnect System

A schematic diagram of a typical[7] three-level VLSI CMOS multilevel interconnect structure is shown in Figure 2.1, where several features of the multilevel interconnect system are shown. In this diagram, the source/drain and polysilicon regions are silicided for the reduction of contact resistance. Some processes use local interconnect (LI) or straps that may be formed as part of the silicide process to reduce device parasitics.[8] Each dielectric layer is planarized to improve lithography and metal step coverage. Contacts and vias are opened in the dielectric films and filled with tungsten, W, as shown in Figure 2.1. The metal layers consist of a stack of films as each layer has a specific application since no single layer can effectively satisfy all the stringent requirements for interconnect metallization. A barrier metal layer (BM) is needed to prevent the aluminum (Al) from interacting with adjacent layers. The main current-carrying conductor that is widely used is an Al alloy and several newer materials are in various stages of development. To reduce reflectivity of the metal layers, an antireflective material (ARM) is used. If a such a layer is not used, the highly reflective metal surface will distort the ultraviolet (UV)

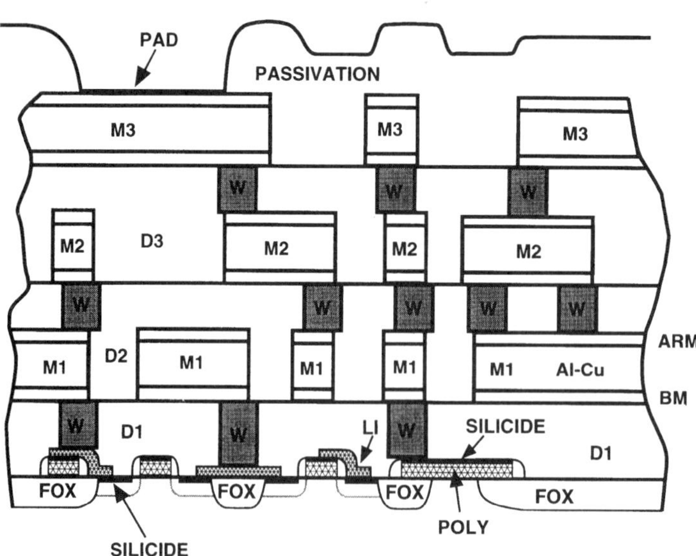


**Figure 2.1** Schematic representation of a multilevel interconnect system.[7] (*© 1989 IEEE.*)

**TABLE 2.1 Terminology of a Triple-Level Metal Interconnect**

| Type of opening | Dielectric | Connecting metallization | Line metallization | Connecting layers |
|---|---|---|---|---|
| Contact | D 1/ILD 1 | Plug 1 | M1 | Poly Si/M1 |
| Via 1 | D 2/ILD 2 | Plug 2 | M2 | M1/M2 |
| Via 2 | D 3/ILD 3 | Plug 3 | M3 | M2/M3 |

light during resist exposure. This distortion leads to bridging resist and poor critical dimension control.

It is a good practice to properly designate layers in a multilevel interconnect system. The whole purpose of the naming convention is to make the process flow convenient for manufacturing. Since there is no industrywide standard nomenclature for multilevel system designation, Table 2.1 shows how to designate various layers of a three-level metal system. This process can be extended to whatever depth and to as many levels as required. Other examples of multilevel interconnect layer designations are available in the literature.[18,19]

The first interlevel dielectric is between polysilicon and metal 1 and is usually a doped oxide. For dielectric layers, the designation could be D1, D2, etc. or ILD1, ILD2, etc., where D refers to the dielectric layer and ILD refers to the interlevel dielectric layer. In cases where more than one planarization is needed, planarization levels may be defined as Planar 1 or 2, etc. The openings in the dielectric layer from metal 1 to polysilicon and diffusion are referred to as contacts. Holes made through postmetal 1 dielectric layers are referred to as vias. In Figure 2.1, W plugs are used to fill contact/via openings and are usually referred to as plug 1, 2, etc. Designating the metal layers that are used for interconnection as M1, M2, etc. is obvious. The metal layer consists of a stack of different metals, such as a barrier layer, a shunt layer, a main conducting layer, and finally an antireflective layer.

## 2.3 Multilevel Interconnect Flow

The flow chart in Figure 2.2 shows the sequence of fabrication operations needed to manufacture an integrated circuit. This flow chart is the basis for discussion in this chapter and is used frequently. Specifically, the discussion will focus on the multilevel interconnect steps from dielectric deposition to metallization.

The first step in the fabrication is the deposition of a dielectric layer that isolates one level of metallization from the next. Since topography influences the step coverage of metallization layers, these dielectric layers are planarized. Once the necessary degree of planarization for a particular level of interconnection is obtained, the next step is to form contacts or vias by etching the dielectric to the

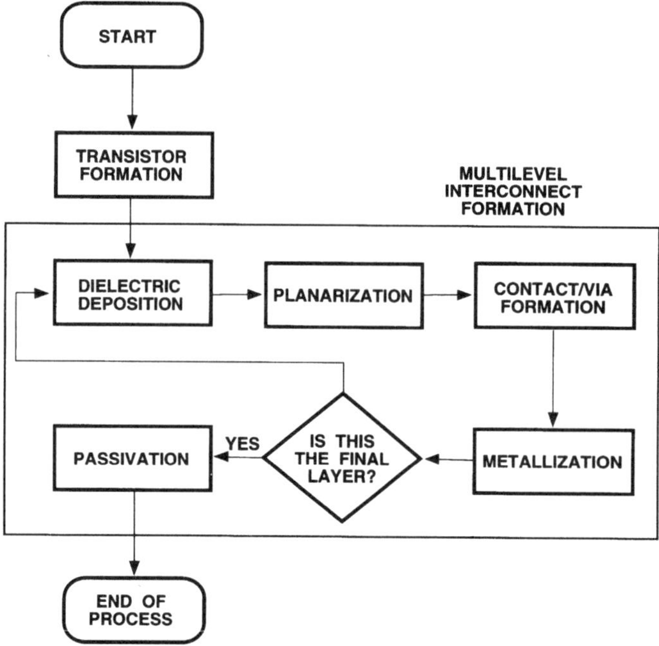


**Figure 2.2** Processing sequence for multilevel interconnects.

underlying substrate. After this operation, a metallization step is used to fill the contacts/vias with appropriate metals for a proper electrical connection to the previous layer. Then metal interconnects are defined. This generic flow of events is repeated as many times as needed for fabricating devices with multiple levels of interconnects.

## 2.4 Dielectric Deposition

In the manufacture of integrated circuits, chemical vapor deposition (CVD) dielectric deposition is an extremely important function.[20–24] Dielectric isolation is essential to separate conducting layers from each other, and this isolation becomes more and more critical as the device geometries shrink and multiple layers of interconnects are used. In addition to isolation, dielectrics are used as a protective overcoat on the fully processed device. This protective overcoat protects the device from external contamination and scratches.

CVD dielectrics are used at different levels in integrated circuit (IC) fabrication processes. The deposition process varies depending on the thermal sensitivities of underlying films and overall thermal budget for the device. Film properties largely depend on the CVD system employed and the chemistry used for deposition. There are three categories of CVD systems commonly available commercially.[22] These are the low-pressure CVD (LPCVD), atmospheric pressure CVD (APCVD), and plasma-enhanced CVD (PECVD).

At the transistor level, CVD-deposited oxides and nitrides are used as sidewall spacers in lightly doped drain (LDD) structures. LDD structures are critical in metal-oxide semiconductor (MOS) devices as they minimize the field strength close to the gate edge, thereby inhibiting injection of hot carriers. Once transistors are formed, an isolation layer is deposited that is usually an oxide doped with boron (B) and/or phosphorous (P). Contacts to polysilicon and source/drain areas are formed in this film and filled with a conducting layer. Premetal dielectric deposition takes place at high temperatures, unlike the postmetal CVD oxides. Postmetallization CVD dielectric deposition offers a new set of challenges, as they have to be deposited at lower temperatures (< 400°C). These films are usually silane or tetraethylorthosilicate (TEOS)–based oxides and sometimes polyimides are used as interlevel dielectrics (ILD). These ILD layers are subject to various planarization schemes in order to minimize the impact of topography on the next level of interconnects. The final protective overcoat or passivation layer is usually a combination of silicon nitride or oxynitride, oxide, and polyimide films.

### 2.4.1 Applications

In very large-scale integration/ultra-large-scale integration (VLSI/ULSI) technologies dielectric materials are used for various applications. Some of these are the following.

1. For isolation of active areas in circuits
2. Isolation of multiple interconnect levels
3. Protective overcoat against moisture/impurities
4. Sacrificial layer for planarization
5. Alkali ion gettering
6. Stress compensation
7. Alpha particle shielding
8. Transistor gates
9. Etch mask (e.g., for trench etching)

### 2.4.2 Cause and effect analysis

The cause and effect diagram in Figure 2.3 illustrates the various influencing factors that effect the properties of the dielectric films. There are four basic factors that need to be considered when selecting and characterizing a dielectric film for use either as an interlevel

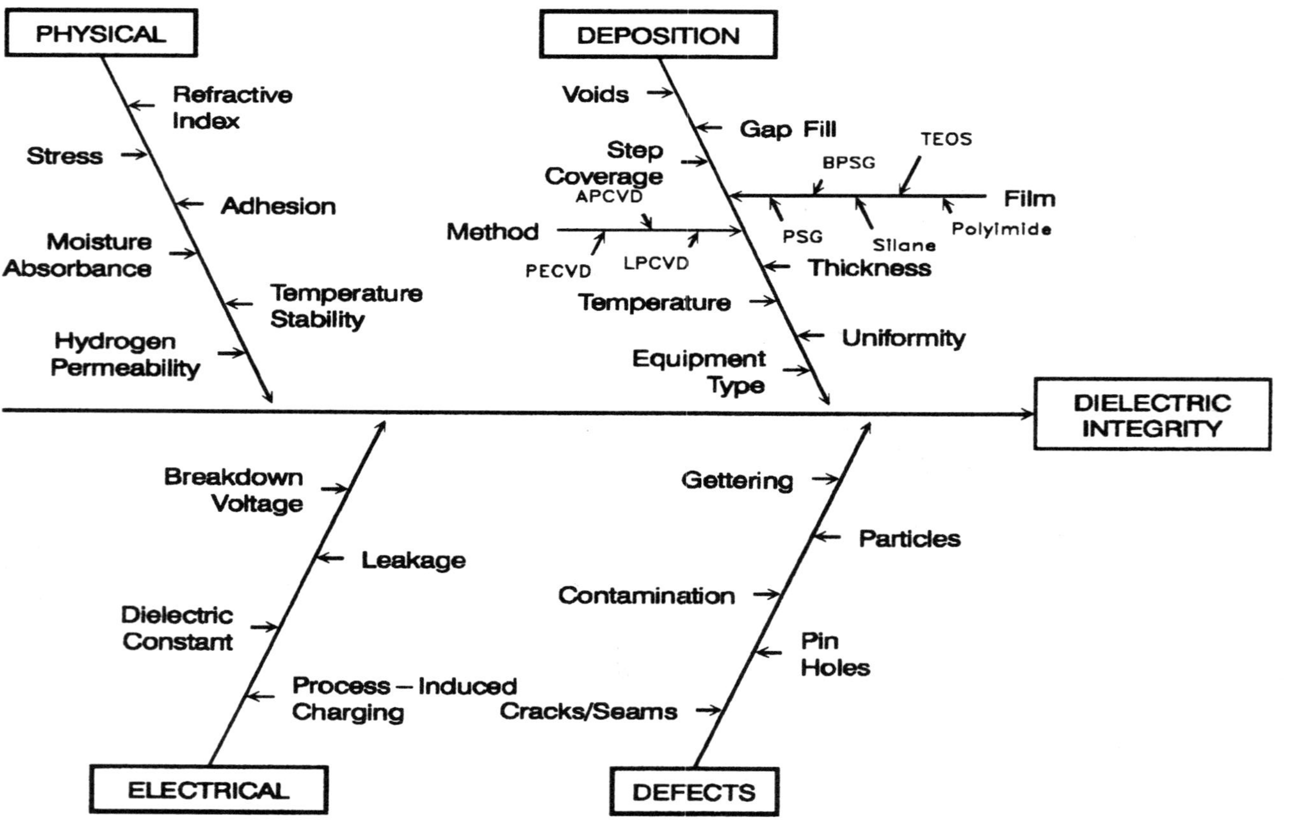


**Figure 2.3** Cause and effect diagram for dielectric integrity.

dielectric layer or as a passivation layer. These are given below and each forms a separate branch in the cause and effect diagram.

**Physical properties/defects.** Physical properties of the dielectric layer are very important as they influence the films below and above it. Compared to thermal oxides, CVD oxides are inferior in quality and integrity. They exhibit higher levels of defects that may be particles, pin holes, and/or contamination. The deposition parameters determine various physical and electrical properties of the film. The refractive index of a dielectric is a good monitor to evaluate the quality of the film. For example, the refractive index of silicon dioxide grown thermally is 1.46. A higher refractive index suggests a Si rich film and a lower value indicates a porous film. Porous films absorb moisture which can cause damaging transistor threshold voltage shifts. A comparison of refractive index of the CVD oxide to the thermal oxide gives an indication of its relative quality.

**Deposition.** CVD deposition basically consists of introducing reactant gases, like silane ($SiH_4$) and oxygen ($O_2$), along with inert gases (for dilution) into the reaction chamber. In the reaction chamber wafers are arranged for uniform exposure to reactant gases. Once the energy (thermal or plasma) to initiate and sustain the reaction is supplied, reactant gases chemically react to form the dielectric film. Reactions can take place either on wafers or close to them. This results in uniform films with good quality.

In order for the CVD film to isolate any two conducting layers, it must be able to cover steps on which it is deposited. Problems in step coverage arise as metal spacings shrink and metal thicknesses increase with each subsequent interconnect layer. The following analysis is valid for any film, dielectric or metal, being deposited over a step.

Figure 2.4 explains what step coverage is in relation to film thickness and Equations (2.1) through (2.3) show the relationship between

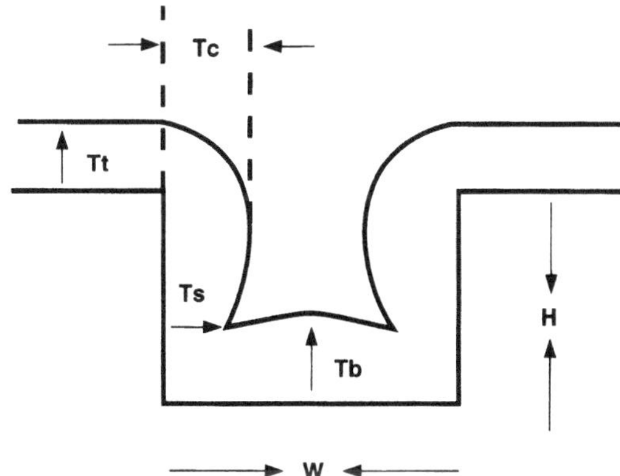


**Figure 2.4** Step coverage analysis.

step coverage and film thickness. Step coverage at a particular location is simply the ratio of the thickness at the location to the nominal film thickness.[19,25] The aspect ratio is the ratio of the height to the width of the gap that must be filled. The higher the aspect ratio, the greater the difficulty in filling the gap.

$$SC_S = \frac{T_S}{T_t} \times 100 \tag{2.1}$$

$$SC_b = \frac{T_b}{T_t} \times 100 \tag{2.2}$$

$$\mathrm{AR} = \frac{H}{W} \tag{2.3}$$

where $SC_b$ = bottom step coverage
$SC_s$ = side step coverage
$T_s$ = thickness (thinnest point)
$T_b$ = thickness (bottom)
$T_c$ = cusp thickness (top)
$T_t$ = thickness on flat surface
AR = aspect ratio ($H/W$)
$H$ = height
$W$ = width

A model[26] has been proposed that explains the deposition and step coverage of plasma-enhanced CVD (PECVD) TEOS films. This model assumes that activated atomic oxygen is responsible for silicon dioxide deposition. The explanation is that two reactions take place within the plasma that modulates the concentration of the atomic oxygen. Atomic oxygen is generated by ionization of oxygen molecules and depleted through volume recombination and chemical reaction on the surface. Table 2.2 summarizes the factors that influence step coverage and deposition rate. Optimum conditions should be determined by factorial experiments that offer a compromise solution between high deposition rate and increased step coverage.

**TABLE 2.2 Factors Influencing Step Coverage and Deposition Rate**

| Parameter | Step coverage | Deposition rate |
|---|---|---|
| RF power ↑ | ↓ | ↑ |
| Pressure ↑ | ↓ | ↑ |
| Temperature ↑ | ↓ | ↓ |

↑ = increase; ↓ = decrease.

Another important CVD film deposition concern is the ability of the film to fill small gaps without voids. Gap fill becomes an issue as the metal pitches become smaller (i.e., aspect ratios become larger). In conformal CVD deposition, cusping usually occurs as the film deposition takes place simultaneously from both sides of the trench, as shown in Figure 2.4. The cusping of the film over steps can be defined as follows:

$$C = \frac{T_C - T_S}{T_S} \times 100 \tag{2.4}$$

where $C$ = percentage of cusp
$T_c$ = thickness (cusp)
$T_s$ = thickness (side)

With continued deposition the leading edges of the cusp meet and form a void as further deposition takes place over the void. The voids may cause problems during planarization and etching as these voided regions are highly stressed and trap moisture/solvents that could outgas under high vacuum and temperature. In order to eliminate the void formation, some commercially available CVD systems employ a deposition/etch process. After depositing a layer of oxide, the film is etched back and then the dielectric is redeposited. The etchback process causes a spacer formation between the steps in the gaps, and reduces the aspect ratio. This mechanism is further discussed in the TEOS oxides section below. Further depositions and etches eliminate the voids. This is required for submicron gap filling as the aspect ratios increase with downward scaling of device geometries. A schematic representation of void formation in the dielectric film due to high aspect ratios is shown in Figure 2.5.

**Electrical properties.** Understanding the electrical properties of the film is very important because they dictate the compatibility of the dielectric layer with the conducting films and also with other CVD films. The main purpose of the dielectric layer is to isolate active areas and conducting layers from one another. Although the dielectric layers do not conduct any current because of their insulating nature, they can modulate the electrical performance of the integrated circuit

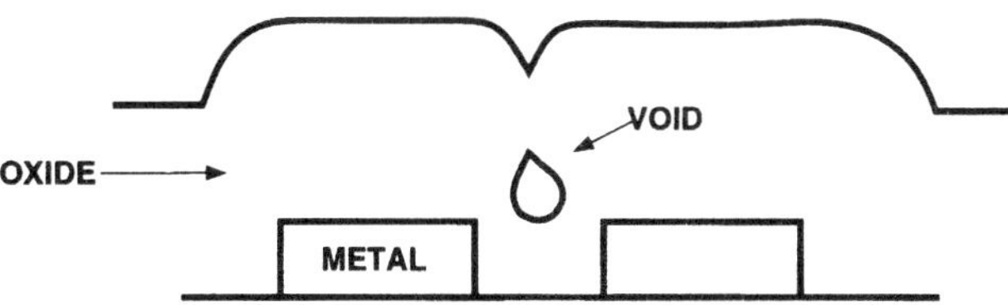


**Figure 2.5** Schematic representation of void formation during dielectric deposition.

if their quality degrades. Dielectric films can induce spurious and unwanted charging on the transistor, which has a detrimental effect on the device reliability.[27–29] The main electrical properties of interest for any dielectric layer are discussed below.

**Capacitance.** As the CVD dielectric film is sandwiched between two conducting layers, the capacitance created modulates the switching speeds of the devices, as explained in Chapter 1. Voids that are formed as a result of the deposition process in narrow spaces could alter the dielectric constant of the insulator between the metal lines as they may trap moisture and other contaminants. The capacitance is a function of the thickness and dielectric constant of the film and both these parameters are considered when designing interconnect layouts. The following equation expresses the relationship between the capacitance of the insulator to its thickness and dielectric constant

$$T_{ox} = \frac{\epsilon A}{C} \tag{2.5}$$

where $T_{ox}$ = thickness of the oxide
$\epsilon$ = dielectric constant of the oxide
$A$ = area of the capacitor
$C$ = capacitance

In order to ensure good dielectric integrity, films with high dielectric constants are desirable.

**Breakdown voltage/leakage.** The breakdown voltage of an insulator is a good measure of its quality and its strength in withstanding an electric potential difference across it. Typical[30] breakdown voltages for insulating films should be greater than 5 MV/cm. Oxide breakdown is an irreversible destruction of the dielectric. However, leakage or tunneling may not necessarily be destructive.

**Process-induced charging.** Process-induced charging is a problem that is not only affected by ion implantation and plasma etching/cleans, but also by plasma enhanced PECVD.[27–29] In PECVD, energy to initiate the chemical reaction of reactant gases is primarily due to the ionization of the gases in a plasma. These ions can induce charging in the gate regions with high direct current (DC) bias, if the gates are floating and do not have adequate charge dissipating structures/mechanisms in place. The charging is a function of the magnitude of the RF power (increased ion bombardment) and its ramp rate as the power modulates the plasma potential and the energy of the ions. If deposition is started with a high radio frequency (RF) power value, the process charging susceptibility of the device is higher than a slow, steady increase in RF power. In addition to this, some commercially available systems employ a sequence of deposition and etch to improve gap fill properties of the CVD film. The etch portion of this sequence may also contribute to the problem of charging. The correlation between power ramp and charging is still not well understood and needs to be examined more closely. A more detailed examination of process induced charging and associated hot carrier degradation is given in Chapter 4.

### 2.4.3 Pre–metal 1 dielectric BPSG[31–43]

A dielectric layer is needed between polysilicon and the first metal for insulating polysilicon interconnects from metal interconnects. Current technologies use a metal silicide to lower contact resistance and these films are susceptible to degradation under high temperature conditions. Therefore, borophosphosilicate glass (BPSG) processing and its reflow must not degrade the silicide layers or alter the junction depths of the transistors.[32,37] In order to satisfy this requirement, BPSG has replaced phophosilicate glass (PSG) as an interlevel dielectric material. PSG films require high temperatures of around 1000°C for reflow and also require a high concentration of phosphorous to improve the flow. The high phosphorus concentrations cause a whole new set of problems, for example, metal corrosion. To decrease the reflow temperature of the PSG films, the oxide is doped with a boron compound. Lowering of the temperature is achieved by addition of boron which makes BPSG quite compatible with the current submicron technologies. The primary function of phosphorous in BPSG is to protect transistors from mobile ion contamination such as sodium. It has been shown[43] that BPSG films getter (or trap) $Na^+$ and other mobile ion contaminants, similar to PSG films.

**Deposition.** BPSG films can be deposited by CVD processes that are based on atmospheric pressure CVD (APCVD), low-pressure CVD (LPCVD), and PECVD techniques at temperatures less than 450°C. BPSG is an amorphous film that is formed by reacting silane ($SiH_4$), phosphine ($PH_3$), diborane ($B_2H_6$), or boron trichloride ($BCl_3$) and $O_2$ diluted with an inert gas, usually nitrogen ($N_2$), at the surface of the wafer. Control and management of gas flow ratios and deposition conditions are important for obtaining the desired film thickness and dopant concentrations. BPSG is formed by the combined oxidation of the various hydrides. The oxidation kinetics are different for BPSG as compared to the oxidation kinetics of any single hydride. Table 2.3 summarizes the impact of varying several process parameters such as flow rates and dopant concentrations, etc. on the weight percent of Boron and Phosphorous and the film deposition rate. Nitrogen is used as a diluent for the hydrides. In the case of APCVD BPSG deposition

**TABLE 2.3 Interrelationship between BPSG Deposition Parameters**

| Parameters | Wt. % B | Wt. % P | Deposition rate |
|---|---|---|---|
| $[B_2H_6]$ ↑ | ↑↑ | ↓↓ | — |
| $[PH_3]$ ↑ | — | ↑↑ | — |
| $O_2$:hydride ↑ | — | — | ↑ |
| $B_2H_6$ flow ↑ | ↑↑ | ↓ | ↑↑ |
| $PH_3$ flow ↑ | ↓ | ↑↑ | ↑ |
| $N_2$ flow ↑ | ↓ | ↓ | ↓ |
| Deposition temperature ↑ | — | — | ↓ |
| Pressure ↑ | ↓ | ↓ | ↓ |

↑ = increase; ↑↑ = large increase; ↓ = decrease; ↓↓ = large decrease; — = no significant impact.

exhaust pressure is one of the key parameters. The greatest impact on B and P concentrations is obtained from modulation of concentration and flow rate of their respective hydrides. Selection of the boron and phosphorous weight percent for a particular device depends on its thermal budget and planarization requirements.

The relationships in Table 2.3 are observed and monitored in a typical BPSG process that is used for device fabrication.[40] When the $PH_3/SiH_4$ ratio increases, the phosphorous concentration in the film also increases, but has little effect on the boron concentration. However, the interaction is different when the $B_2H_6/SiH_4$ ratio is increased. This causes a significant increase in the B concentration, but also causes the P concentration to decrease. The reason for this may be explained by the reaction kinetics of the oxidation of the hydrides. The oxidation of the boron hydride proceeds much faster than oxidation reactions of both silane and phosphine as diborane reacts more readily with oxygen. The oxidation reactions[42] of the three hydrides are as follows:

$$B_2H_6 + 6N_2O \rightarrow B_2O_3 + 3H_2O + 6N_2 \tag{2.6}$$

$$SiH_4 + 4N_2O \rightarrow SiO_2 + 2H_2O + 4N_2 \tag{2.7}$$

$$aPH_3 + bN_2O \rightarrow cP_2O_3 + dP_2O_5 + eH_2O + fN_2 \tag{2.8}$$

where coefficients $a$ though $f$ are unknown.

**Densification.** As the contacts are defined in the BPSG film, the film integrity and flow characteristics are critical for contact lithography and etching. In 1 μm and larger technologies, the densification is done following the deposition of film and then the contacts are defined. Reflow of contacts after contact formation changes the profiles of the contacts and helps improve the metal step coverage. In submicron technologies the control of contact dimensions becomes difficult if reflow is done after the contacts are formed. These technologies rely on other contact planarization techniques and not on BPSG reflow process. Therefore, the densification and reflow may be combined and performed in one single step.

During oxidation of the hydrides of B and P their respective oxides are formed. The as-deposited film contains these oxides and is amorphous in nature. The densification process must take place soon after deposition otherwise these oxides would react with the moisture in the ambient and form crystals. The densification of the BPSG film in steam at around 800°C causes hydrolysis of the oxides and also decomposition and vaporization of some of them.

In the BPSG film, P is present as $P_2O_5$ and $P_2O_3$ oxides.[42] Hydrolysis of these oxides results in formation of their respective acids. Boron oxide also reacts with water to form boric acid. This reaction can take place at room temperature and may result in boric acid crystal formation. This reaction is given below.

$$B_2O_3 + 3H_2O \rightarrow 2H_3BO_3 \tag{2.9}$$

Acids formed by hydrolysis of both boron and phosphorous oxides react further and may cause formation of $BPO_4$. The impact of these byproducts on the device defect density is discussed below.

**Reflow.** The purpose of reflow is to planarize the BPSG film so that underlying topography may be transformed to a gentle and smooth surface. The BPSG flow angle decreases as the reflow temperature increases. However, the reflow temperature is constrained by the device channel length specifications as an increase in temperature may alter the source/drain profiles. In addition, silicided polysilicon and source/drain areas are sensitive to increases in temperature as the silicide will become unstable.

It has been reported[42] that for a given BPSG composition, achieving equivalent flow angles in both nitrogen and steam ambients requires a 70°C or higher temperature in the dry nitrogen ambient. The steam ambient leaches the dopants from the top surface of the BPSG film. This has beneficial effects as the subsequent metal films would contact a lower doped oxide. Steam also oxidizes the interface of BPSG and silicon substrate to form a thin layer of oxide that prevents auto doping of the wells.

**Flow and defect characteristics.** Success of the reflow of the BPSG and its impact on planarization is measured by the flow angle. Flow angles are not only dependent on film deposition conditions and reflow parameters, but also on the underlying topography. BPSG film thick-

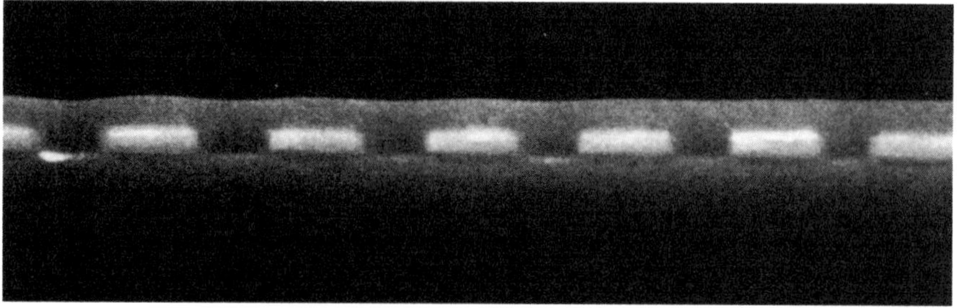

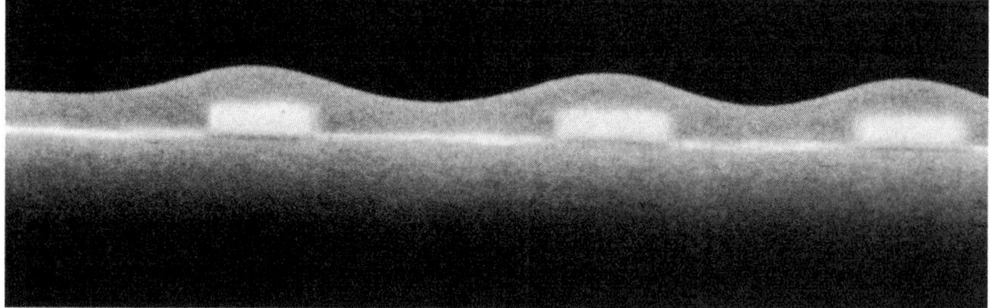

**Figure 2.6** Pre-metal 1 planarization—BPSG over topography.

ness tends to be larger over dense topography as compared with isolated structures. The reflow process contributes to thickness variation by making the BPSG flow which smooths the topography by depletion of the film over steep structures and accumulation in the gaps between polysilicon interconnects. This presents a situation of differential contact depths during the contact etch process because the thickness of BPSG varies over dense and isolated structures. The scanning electron microscope (SEM) micrograph in Figure 2.6 illustrates the flow characteristics of BPSG over steps with varying spacing.

To improve the planarization and obtain lower flow angles, boron and phosphorous concentrations need to be increased. However, this increase may have a detrimental effect on yield and reliability of the device due to the formation of BPSG crystals. There are two types of crystals that are specific to BPSG processing. These are boric acid and borophosphate crystals. Boric acid crystals form at room temperature and are quite easily soluble in water and are usually removed in the steam densification/reflow cycle. In order to control the formation of crystals, the time delay between the BPSG deposition and densification/reflow must be minimized. The boric acid crystals are quite large, between 5 and 10 μm in size. The second type of defect is formation of the borophosphate ($BPO_4$) crystals which are not water soluble. These crystals are formed during steam reflow when the reflow parameters are less than optimal. $BPO_4$ crystals are small in size, usually less than 1 μm.

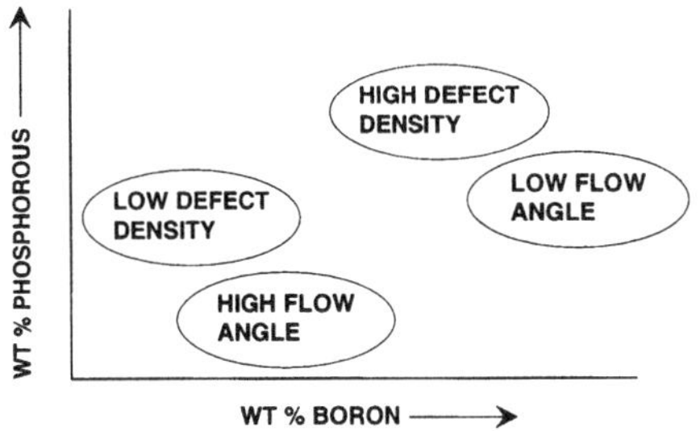


**Figure 2.7** Impact of B and P on defect density and flow angle of BPSG films.

In order to minimize crystal formation lower concentration of dopants is preferred, but such action may lead to higher flow angles. The relationship between the dopant 0 concentration, flow angles, and crystal formation is illustrated in Figure 2.7. This presents an interesting problem—lower flow angles are needed to enhance the planarization but at the same time lower defect densities (due to crystal formation) are needed to prevent yield degradation. Such constraints force this process to operate within a very narrow process window.

### 2.4.4 Post–metal 1 dielectric CVD oxides

The post–metal 1 dielectric deposition process comes with its own stringent requirements for submicron devices. Note that for post–metal 2 or 3, etc. similar process methodology is used; the dielectric thickness may vary from layer to layer due to metal thickness differences between one level and the next.

CVD technology that has been used up to about 1.0 μm design rules has been based on the oxidation of silane.[44] Silane oxide does not deposit conformally over underlying topography and hence is not adequate in filling gaps, of aspect ratios of 0.5 or larger, for submicron technologies.[63] TEOS-based oxides have improved the conformality of the deposited oxides. Improved conformality helps in preventing voids in thick films and also presents a quasiplanarized surface. But is TEOS deposition process adequate for sub-half-micron technologies? This section examines the relative advantages of TEOS-based oxides over silane oxides and process integration issues involved in incorporating a dielectric film that satisfies both the physical and electrical requirements for submicron devices.

**Silane oxides.** Oxides formed by the oxidation of silane have been limited in their use because of their deposition characteristics, especially for sub-micron process technologies with high aspect ratios.[44–47] Silicon dioxide is formed by the reaction of $SiH_4$ and $O_2$ in a low pressure regime of less than 3 torr and around 400°C. This reaction takes place at the wafer surface and is primarily governed by the rate of arrival at the surface of the reactant species from the gas phase and of course the temperature of the surface.[44] To achieve a conformal coverage of steps or fill gaps, the surface mobility of the active species must be high and the reaction must be uniform at the surface. It has been observed[44] that during the deposition of silane-based oxides, the active species have very little surface mobility, possibly less that 1 μm. The surface mobility could be enhanced by increasing the wafer temperature, but this poses a major problem with the metal layers underneath the interlevel dielectric. In order to prevent the aluminum interconnects from voiding and to suppress hillock formation, the temperature must be low.

Since silane-based oxides suffer from poor surface mobility, the deposition of the oxide in narrow gaps and over wide structures varies and is a function of the arrival angle of the reactive species from the gas phase. In wide-open areas, the arrival angle is high and therefore the deposition rate is also high. On flat wafers the deposited oxide is quite uniform. However, in narrow spaces the arrival angle is a function of the aspect ratio. The higher the aspect ratio, the smaller is the arrival angle. The amount of reactant species arriving and reacting to form $SiO_2$ in high-aspect ratio spaces is less over open and wide structures. This causes localized differential deposition rates, probably due to gas depletion. Therefore, for a given film thickness, the open areas will reach the required thickness before the narrow gaps. As the deposition continues over time, the film gets thicker and voids begin to form in narrow spaces. The void formation is due to poor surface mobility and deposition rate variations.

Silane chemistry is adequate for technologies above 1 μm. For submicron technologies this technique of ILD deposition may not be adequate if used by itself. Silane chemistry may be adapted for submicron technologies by using a multilayered dielectric layer in combination with an appropriate planarization process. For example, silane-based oxides in combination with spin-on-glass (SOG) have been used quite successfully.[85] A thin layer of oxide deposition followed by SOG can be either used in an etchback or a nonetchback mode. Details on SOG planarization are given in the appropriate section. Such a dielectric stack allows sufficiently thin oxide film to be deposited so the problem of void formation is eliminated. Gap fill is achieved by the SOG film. From a safety point of view, silane gas delivery and exhaust management is very important as silane is very flammable.

**TEOS oxides.** TEOS-based oxides[48–59] deposited at low temperatures (<400°C) have been developed as an alternative dielectric to silane-based oxide because they tend to have better gap fill properties.[45] Here again, the surface mobility is believed to play a pivotal role in enhancing the step coverage and also the gap fill. The reactant species in TEOS-based oxides tend to have a higher surface mobility than the silane-based oxides. In addition to a higher mobility, the TEOS reactant species exhibit a decrease in sticking coefficients.

TEOS silicon dioxides may be deposited in a plasma-enhanced mode where the TEOS reacts with oxygen in a plasma. This type of oxide may be designated as plasma-enhanced TEOS (PETEOS). PETEOS silicon dioxide has better conformality than silane based oxide, but it is not perfect. Typical conformality values are less than 75 percent in narrow spaces.[63] Silicon dioxide may also be formed by LPCVD reaction of TEOS with ozone at temperatures of less than 400°C. Oxide formed by this technique is termed thermal TEOS (TTEOS). The TTEOS reaction proceeds as follows, even though this reaction is a result of several intermediate reactions:

$$Si\text{-}(O\text{-}C_2H_5)_4 + O_3 \rightarrow SiO_2 + \text{by-products} \tag{2.10}$$

TTEOS silicon dioxide films are nearly 100 percent conformal but suffer from being very porous and absorbing a significant amount of moisture.[57] To further enhance the step coverage and film integrity of TTEOS oxides, the deposition pressure may be varied. TTEOS/$O_3$ may be reacted at atmospheric or sub-atmospheric pressures. The change in pressure causes the films to be denser and absorb less moisture. For the 16-megabyte (MB) DRAM process a subatmospheric (600 torr) TTEOS deposition has been used to fill trenches 0.5 μm wide by 1.0 μm deep.[58]

The current strategy to fill narrow spaces without voids is to use TEOS-based oxides and simultaneously decrease the aspect ratio of the gaps so that the dielectric material can fill the spaces easily. But how can we decrease the aspect ratio? A deposition/etch method[59] may be used to improve the deposition characteristics using TEOS based oxides. Depositing a specified thickness of oxide and then etching it back partially creates a spacer in the gaps. The spacer changes the aspect ratio and further deposition of oxide can be done without void formation. The spacer formation may be achieved by either argon (Ar) sputter etching of the oxide and/or reactive ion etch (RIE) etchback of the oxide. Deposition, etchback, and redeposition are all done in one machine, but in different chambers. Generally the PETEOS and TTEOS are used in a deposition/etchback dielectric deposition scheme where the TTEOS is used only for gap filling.

After the TTEOS is used to fill the gaps, through single or multiple depositions and etchbacks (Ar sputter etching or RIE), PETEOS oxide

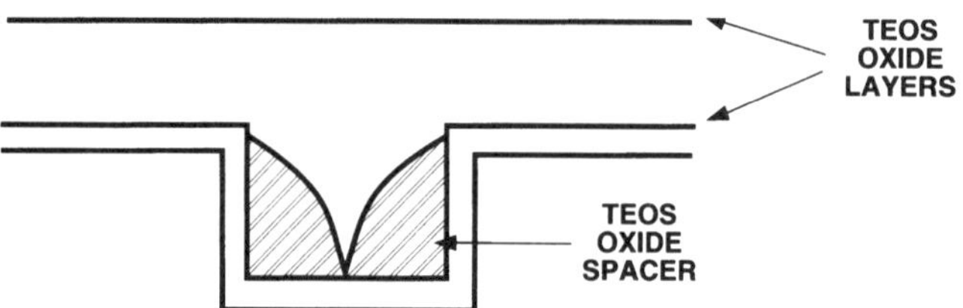


**Figure 2.8** TEOS-based oxide deposition in narrow spaces.

or other oxides are redeposited for planarization. Figure 2.8 shows a schematic illustration of the TTEOS spacer sandwiched between two PETEOS layers.

PETEOS is a denser and dryer film than TTEOS because the reaction of TEOS and oxygen in a plasma is more efficient in dissociation and enhancement of the reactant species in this gas phase oxidation. Furthermore, the plasma assists the surface reactions through ion bombardment. This results in PETEOS films having higher deposition rates, lower hydrogen concentration (therefore a dryer film) improved bonding structure (denser film), and overall better film properties than TTEOS. One of the drawbacks of using TEOS-based oxides is their effect on the transistor threshold voltage stability. Threshold voltage shifts can occur when moisture diffuses from the dielectric layer to the gate region where it creates electron traps.

## 2.5 Dielectric Planarization

This section deals with planarization of dielectric layers and follows the dielectric deposition module as shown in Figure 2.2. Planarization has become an integral part of the multilevel interconnect scheme where topographical interferences with fine geometry patterning and etching pose a problem.[60–64] The source of this problem lies in the fact that several levels of metal interconnects are required to fabricate integrated circuits with millions of transistors per chip. The current distribution requirements for the different metal interconnects demand appropriate film thicknesses for both the metal and the dielectric films between them. A stack of several layers of metal separated by insulating films make the topography very uneven with extremely large steps. Planarization is essential for the following reasons.

1. *Step coverage.* The step coverage of metal impacts the current carrying integrity of the interconnects. Poor metal step coverage will cause current crowding and will lead to electromigration failures. Planarization is required for multiple-level metal technologies and beyond.

2. *Differential bias/critical dimensions.* Planarization, along with the antireflective layers, prevents spurious UV light reflections from the sides of metal as it goes over steps. Such reflections can cause loss in critical dimensions and also cause defects in the metal line. Planarization is required to overcome differential bias due to variations in resist thickness on unplanarized surfaces.

### 2.5.1 Requirements

There are basically two requirements for any planarization process.

1. *Uniformity.* The uniformity is an important output parameter that needs to be characterized and monitored to ensure that all the dielectric is planarized irrespective of the circuit density. One of the key uniformity parameters is the within-a-die (WID) uniformity. This determines if the planarization is local or global in nature, which will be described in this section. Within-a-wafer (WIW) uniformity is also an important parameter as the via etch uniformity is influenced by it.
2. *Manufacturability.* The planarization process has to be manufacturable. This means the machine through put must be reasonably high to support high-volume production. Under the conditions of high output the defect density must be very low as yield and reliability must not degrade.

### 2.5.2 Local and global planarization

Planarization can be local or global in nature.[65,71,74] *Local planarization* refers to planarization in dense regions of the device where planarization[71] occurs over submicron geometries with pitches varying from less than 2 to about 4 μm. *Global planarization* refers to planarization over a large distance of about 150 μm or greater.[71] In some planarization procedures, it is essential to optimize for local planarization and sacrifice global planarization. This is a problem that is usually faced in the etchback type of planarization processes. For example, memory circuits, such as DRAMs, have a very dense concentration of structures in the array as compared to the periphery of the die. Therefore, they need an array-optimized planarization process. However, for microprocessors and other logic devices, planarization should be optimized evenly across the die as the structure layout is not as systematic and symmetrical as in the memory circuits.

### 2.5.3 Summary of planarization technologies

There are several dielectric planarization processes currently being used for 1 μm and submicron process technologies.

**Resist etchback.**[74–77] This is a very common and convenient method for dielectric planarization. Planarization of the oxide surface is achieved by coating the wafer with a sacrificial resist layer and then etching it back with approximately equal oxide and resist etch rates. After resist is spun on the wafer, the surface is planarized because the resist fills the depressions and is thin over steps. During plasma etchback, the oxide over the high points gets etched before the oxide in the spaces. The etching is continued until the desired oxide planarization is obtained.

**SOG (with and without etchback).**[78–95] There are two methods of using SOG for dielectric planarization. The first is a nonetchback process. This involves spinning, baking, and curing the SOG until the desired planarity is obtained. Additional oxide is deposited over the SOG to meet the dielectric thickness specifications. The second method is an etchback process where SOG is used as a sacrificial layer, similar to resist in the resist etchback process.

**Chemical mechanical polish.**[65–73] Planarization is achieved by polishing the dielectric surface on a rotating abrasive pad. This causes the high spots on the dielectric surface to be polished before the low spots. The chemical component is provided by the slurry which has two primary functions. First, it serves as a lubricant and, second, it helps break the $SiO_2$ bonds at the surface.

**Bias sputtered quartz.**[96] The wafers are planarized by sputter deposition and etch of dielectric material from a target onto the wafer. During the sputtering process, the wafers are held at a negative potential relative to the plasma and the bias voltage may be varied to obtain the desired planarization. In this process, sputter etching and deposition are simultaneously taking place. As the material is being deposited, the high points on the wafer surface are sputter etched. This dual action planarizes the surface.

**Thermal flow.** Planarization using thermal flow is restricted to the dielectric layers over polysilicon and before any metal layers are deposited. The oxide is doped with B and P to improve the flow characteristics and to lower the flow temperature. This process was discussed in detail above in Section 2.4.

**Deposition/etch.** Good conformal coverage may be obtained by using a LPCVD/PECVD deposition/etch process. The material is deposited and etched in sequence a few times until the gaps between metal lines are completely filled. In addition to the deposition/etch sequence, the use of TEOS-based oxides has also helped in achieving good gap fill and planarization.

**TABLE 2.4 Comparison of Planarization Schemes**

| Planarization process | Local | Global | Feature size (μm) | Mfg |
|---|---|---|---|---|
| SOG | Yes | Some | 1 | Good |
| Etchback | Yes | Some | 1 | Good |
| CMP | Yes | Yes | <1 | Fair |
| Thermal flow | Yes | No | <1 | Good |
| Deposition/etch | Yes | No | <1 | Good |

Table 2.4 is a comparison of some planarization processes. The properties of the planarization processes that have been compared are the following.

1. *Local/global planarization.* It important to recognize the primary benefit of various processes. Some have superior local planarization properties and other are suited for a global planarization.
2. *Feature size.* The feature size dictates the process technology the planarization step is best suited for.
3. *Manufacturability (Mfg).* This aspect indicates the adaptability of the planarization process in a high-volume manufacturing environment.

### 2.5.4 Degree of planarization

We can evaluate the success of any planarization technique by measuring its degree of planarization and the planarization distance. The degree of planarization, DP, is a measure of the dielectric step coverage after planarization and is defined by Equation (2.11):

$$\mathrm{DP} = 1 - \frac{(SH)_{\mathrm{post}}}{(SII)_{\mathrm{pre}}} \tag{2.11}$$

where DP = degree of planarization
$SH$ = step height

If DP approaches 1, that means the prestep height is very large as compared to the poststep height. In such a situation the planarization is perfect over long ranges. If the pre- and poststep heights were to be equal then the DP will be equal to zero. This means there is no long-range planarization.

Another figure of merit that is indicative of the planarization effectiveness is the planarization distance (PD). Planarization distance is a measure of the wafer smoothness over long distances and is the distance of the step in topography after planarization. A long PD implies that the variations in topography smaller than the PD will be completely planarized. It is not surprising to find PDs up to a few millimeters for a chemical mechanical polish planarization process. As compared to CMP, SOG planarization may yield a PD of about 2 μm.[65] Higher PD values are indicative of global planarization.

Instead of describing each of the different planarization processes, the focus of this section is to examine the issues pertaining to local and global planarization. To illustrate each of these, a planarization method has been selected that best represents either local or global planarization. For local planarization, SOG process with and without etchback is discussed. For global planarization, chemical mechanical polishing is examined.

### 2.5.5 Global planarization (CMP)

Chemical mechanical planarization (CMP) is gaining wide acceptance as it is the only process to date that can offer a better global planarization for VLSI/ULSI devices than the other planarization processes.[65–73] The use of CMP for planarization was first used[68] in combination with RIE to planarize CVD oxide for shallow trench isolation as part of the 16-MB DRAM process. This combination helped in reducing process nonuniformities associated with each process when used independently. There has been a lot of activity in the development of CMP since then. Using chemical mechanical polishing by itself to planarize interlevel dielectric oxides has been reported with very encouraging results.[65] Therefore, the discussion below focuses on planarization with CMP alone and not in combination with other techniques.

The CMP cause and effect diagram shown in Figure 2.9 illustrates the various parameters that effect oxide polishing. The CMP method may be briefly described as a process where the wafers are held face down against a rotating polishing pad. This pad is constantly lubricated by a slurry that consists of colloidal silica at a high pH. It is this mechanical action that enables the oxide to be removed from the surface. The polish process flattens out height differences, since the high areas of topography are removed faster than low areas. When this is accomplished, all traces of the original topography are removed and an acceptable post polish surface is obtained.

The CMP process can planarize both locally and globally over topography with varying pitches. Local planarization is shown in Figure 2.10; global planarization is demonstrated in Figure 2.11.

The two main components of the CMP process that remove the surface oxide are the mechanical and chemical actions. Both these are necessary for the polishing of the oxide surface and are discussed below.

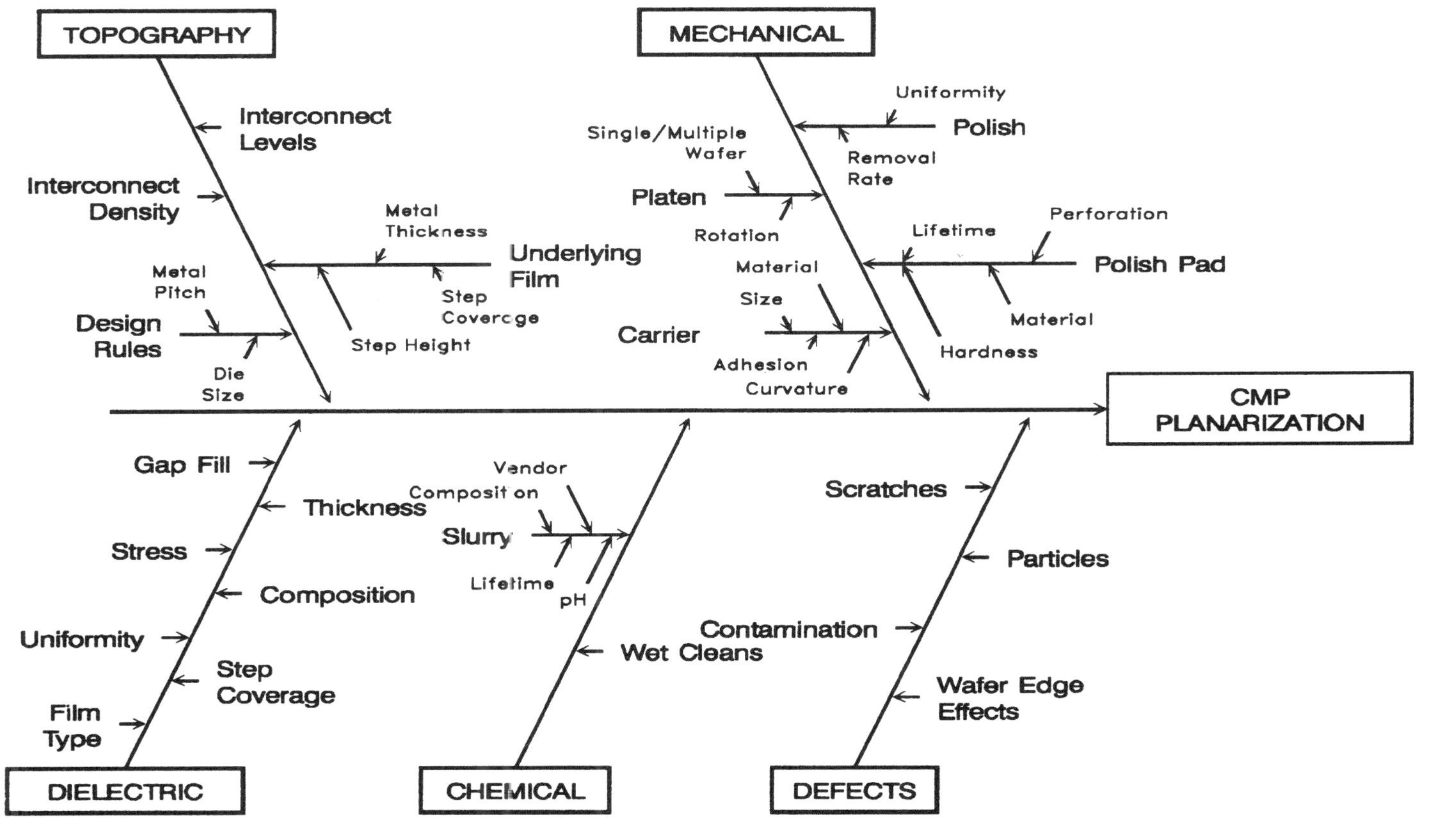


**Figure 2.9** Cause and effect diagram for chemical mechanical planarization process.

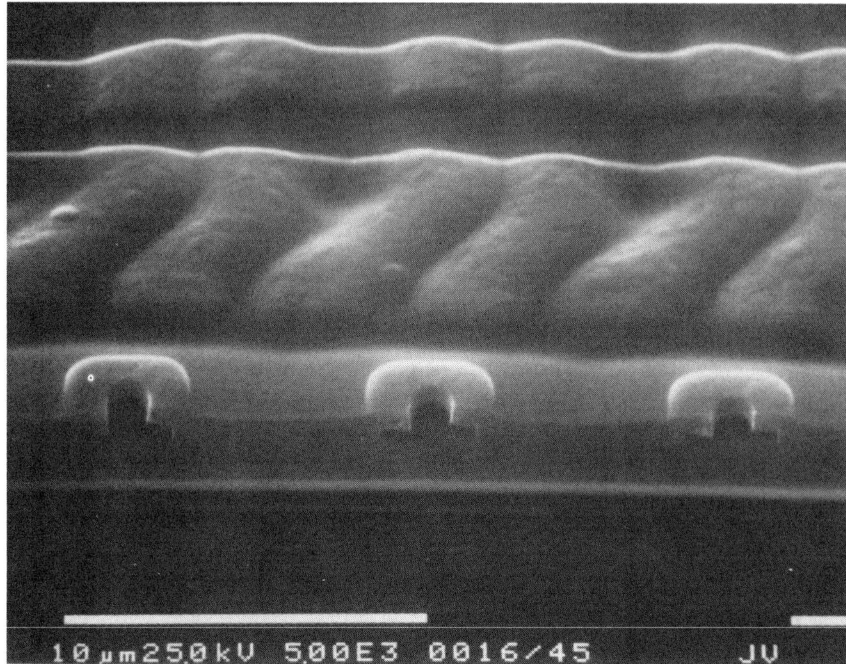


Figure 2.10 Example of local planarization.

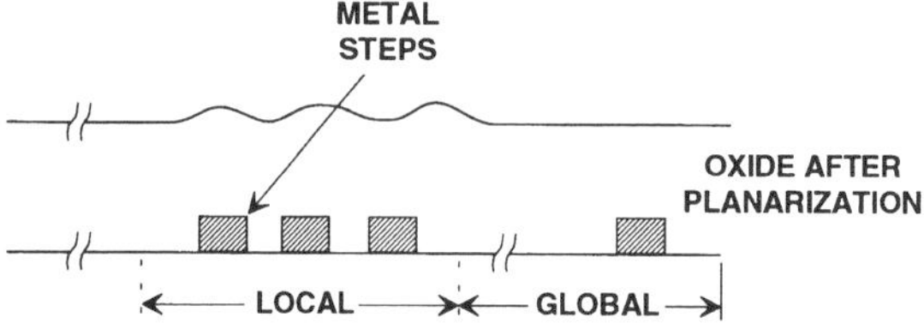


Figure 2.11 Schematic illustration of local and global planarization.

**Mechanical component.** The mechanical action consists of a rotating table or platen on which the polishing polyurethane pad is mounted. The wafer is pressed down against this polishing pad by a wafer carrier, whose shape may be adjusted to improve the polish uniformity. Downward pressure is then applied to the wafer. The hardness and aging of the polish pad effect the oxide planarization. The removal rate of the oxide continuously drops off as the pad ages and the perforations in the pad are blocked and therefore prevent slurry distribution. For effective process control, the pad needs to be conditioned by constant and regular cleaning, usually through an abrasive process

**TABLE 2.5 Process Interactions in CMP**

| Parameter | PD | Removal rate | WIW |
|---|---|---|---|
| Platen speed ↑ | ↓ | ↑ | — |
| Down pressure ↑ | ↑ | ↑ | — |
| Pad age ↑ | ↓ | ↓ | ↓ |
| Pad hardness ↑ | ↓ | ↓ | ↓ |

that opens the channels for the slurry to flow. This action restores the oxide removal rate. The removal rate of the oxide, from a mechanical viewpoint, is a function of the downward pressure, the relative velocity between the oxide surface, and the polishing pad, age/conditioning of the pad, and temperature.

Table 2.5 summarizes the interactions of the parameters that are part of the mechanical component of the polish process. This table shows the effect on the degree of planarization, the removal rate and the WIW uniformity. The up arrow indicates an increase and a down arrow a decrease in value of the parameter. When the effect on the output parameter is negligible it is denoted by a dash (—).

**Chemical component.** The mechanical polishing action is complemented by the chemical component in the form of slurry introduction. The slurry, consisting of suspended colloidal silica in potassium hydroxide, reacts with the dielectric film by forming a hydroxylated surface layer. It is thought that several surface interactions take place between the slurry particle and the oxide surface. These reactions result in the disassociation of the Si-O-Si molecule through the diffusion of water through the oxide. The bond strength between the slurry particle and the wafer surface determines the kinetic coefficient of friction between the two surfaces during polishing. A high pH is needed for the suspension of the silica and for the reaction to take place. The by-products of this reaction are removed by the mechanical action pad and the colloidal silica.

### 2.5.6 Local planarization (SOG)

SOG is being widely used and remains one of the most promising materials for interlevel planarization of VLSI multilevel interconnects.[78–95] By comparison with other planarization techniques, the SOG process is a relatively simple technique for planarizing surfaces with severe topography both below[84] and above the first metal layer. As SOG is similar to $SiO_2$ after its final cure, it can be made a part of the dielectric stack and can be easily integrated with CVD oxides to obtain stable insulating properties. This can be achieved by proper selection of SOG material and optimizing its coating, baking, and curing steps. Unlike CMP, SOG does not offer complete planarization, instead it smooths out the underlying topography. This results in a reduction in via depth differentials. Even though SOG is a simple planarization technique and has several advantages over other processes,[88] the reliability implications of its use must be carefully examined. Moisture-induced transistor threshold voltage instabilities and other reliability issues are some of the major concerns with SOG planarization.[86,87]

SOG is a liquid and thus can easily be spun on wafers. This action causes the SOG to fill gaps or depressions in the topography and provide a planar surface. Several thermal treatments are necessary to ensure that the desired thickness is obtained and the solvents are removed from the SOG. To illustrate the planarization properties of SOG, Figure 2.12 presents a generic relationship between the degree of planarization and the metal line spacing for two different metal thicknesses, h and 2h. In narrow spaces, SOG fills the gaps and is thick in those areas. As the metal line spacing is increased and the gaps widen, SOG thickness in these areas decreases and consequently the degree of planarization decreases. A similar effect is observed when the metal thickness is increased. As the step height is increased, the degree of planarization decreases. The SOG cause and effect diagram given in Figure 2.13 provides an overview of the SOG planarization process. This diagram highlights some of the salient aspects of SOG planarization, which are discussed below. Key to successful implementation of SOG planarization process lies not only on the technique (with or without etchback) used but also on the material.

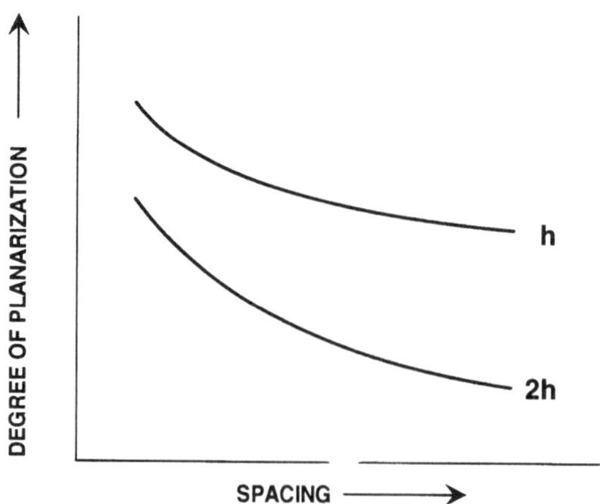


**Figure 2.12** Impact of step spacing and height on degree of planarization.

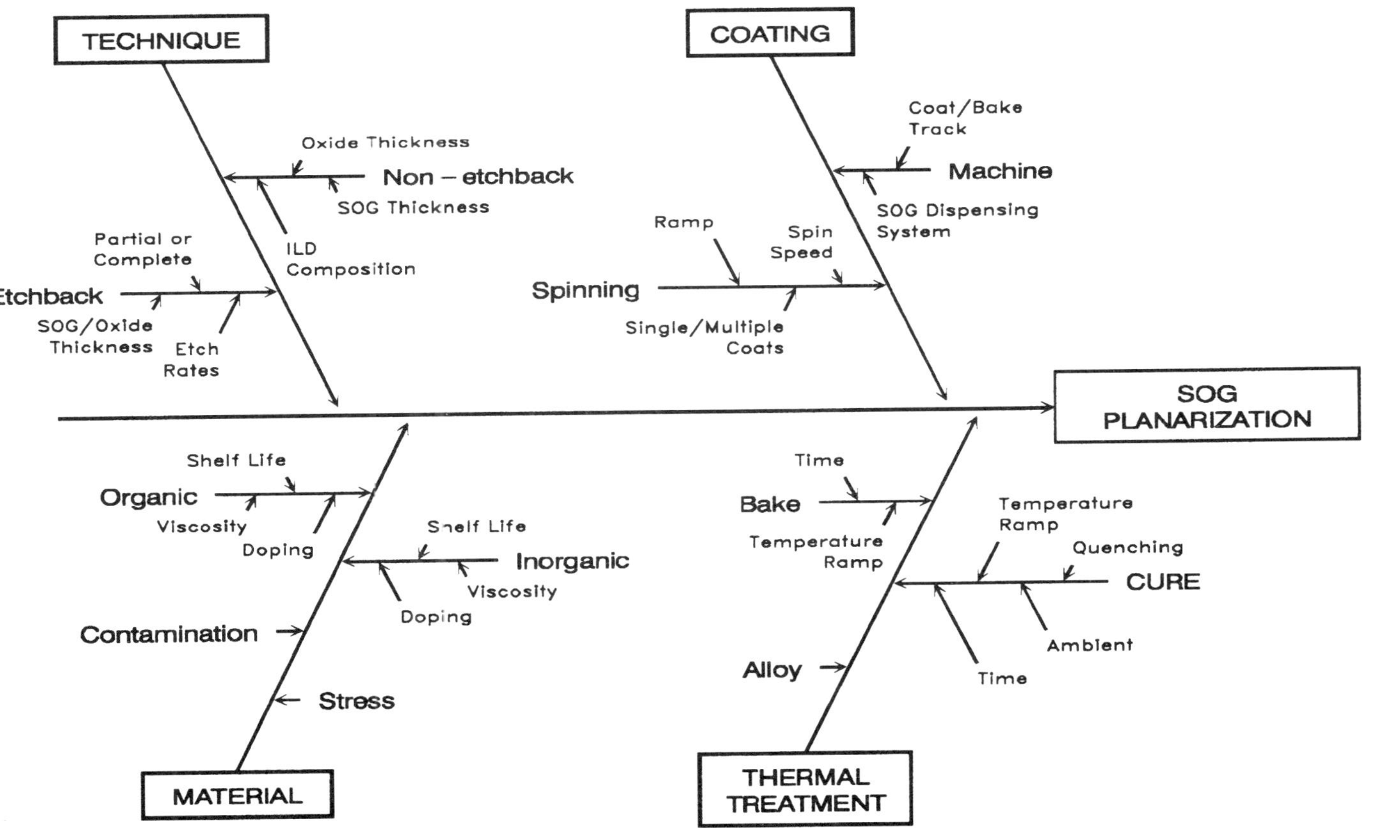


Figure 2.13 Cause and effect diagram for SOG planarization.

**SOG process integration.** SOG integration for planarization above the metal layers has essentially followed two major paths, and each has demonstrated reasonable success in achieving the necessary degree of planarization. These approaches are described below.

**SOG sandwich.** In this technique,[85,93] the SOG is not etched back at all and is sandwiched between two layers of CVD oxide as shown in Figure 2.14. In the sandwich structure the three layers shown have specific purposes that need to be carefully evaluated when using the SOG sandwich approach.

The first insulating layer from the bottom can be either CVD oxide or oxynitride, and it serves as an adhesion and a metal hillock suppression layer that prevents the SOG from coming in contact with the metal. The thickness requirements of this first layer depend on the device design rules. As a rule of thumb, the first layer should at least be able to cover the metal steps and any hillocks that may have formed. The maximum thickness should be approximately less than half the metal spacing to prevent any formation of voids. The middle layer, whose main purpose is to fill gaps, is the SOG. In order to achieve a high degree of planarization, the SOG material selection, processing, and curing conditions must be carefully optimized. The third layer of the dielectric is the final insulating layer, and its minimum thickness is defined by the design rules for isolation and speed for the devices.

If the SOG processing and curing are not optimized, the outgassing of moisture from the SOG through the via side walls may hamper the metal deposition and thus give rise to "poisoned vias." Poisoned vias cause high via resistance or even unreliable vias. The etchback approach ensures that the SOG is removed from all potential via locations so that no poisoned vias can occur. However, this is at the expense of additional process complexity.

**SOG etchback.** This technique[90,91,95] employs either a partial or complete etchback of SOG after it is deposited over the wafers. The methodology is similar to the resist etchback process. In the partial etchback variation of this technique, SOG is spun on the wafers after the first dielectric layer is deposited. The spin parameters are chosen as in the previous case to fill the spaces between the metal lines without any cracking. The SOG is partially etched back, targeting an etch time that leaves some oxide on top of the metal, therefore preventing the metal from being exposed during the etchback process.

During the etchback step, a selectivity of 1:1 is used to etch both the oxide and the SOG at the same rate. A desirable selectivity can be obtained by adjusting the refractive index of the dielectric layer underneath the SOG and also by varying the etch parameters, e.g.,

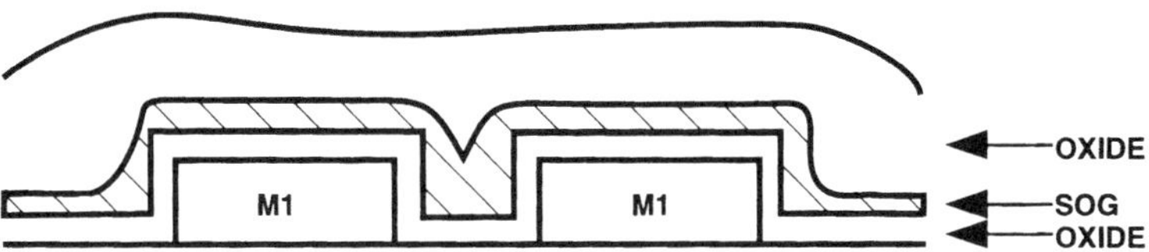


**Figure 2.14** SOG sandwich structure.

the pressure. In case of complete etchback, the SOG is completely etched away. This calls for a thicker base dielectric layer; during the etchback, the oxide and the SOG are etched back at the same rate as long as loading effects are not present. Loading effect causes degradation of planarity due to increase in SOG etch rate as SOG surface decreases. After the SOG etchback, a layer of oxide is again deposited in order to meet the minimum thickness requirements for the insulating film. One of the concerns of this technique is the clearing of SOG in worst-case areas, specifically on top of metal 1 where the polysilicon and the field isolation steps coincide and have minimum space restrictions.

In both the etchback and the sandwich approaches, the key to success is the manipulation of the SOG process parameters to obtain an acceptable planarity. The planarization of SOG may be affected by the planarization of the layer underneath metal 1, which in most cases is a BPSG layer. The flow angle of BPSG after reflow influences the etchback process window. To obtain a larger process window, some device technologies may demand a higher planarity of the BPSG layers than that offered just by reflow process. Postpolysilicon planarization using SOG has been demonstrated to be a viable alternative to the standard reflow processes.

**Material selection.** SOG is available in two different forms: silicates (inorganic) and siloxanes (organic). The chemical behavior of these glasses influences the quality of planarization. Therefore, depending on the SOG planarization requirements, the appropriate material needs to be selected. These SOGs can be doped with B or P to enhance the flow properties of the glass.

The management of the carbon content in SOG is very important for achieving good planarization performance. Increasing the carbon content decreases the film polymerization as the polymerization sites are blocked by methyl and phenyl groups. Inorganic SOG does not have any methyl or phenyl groups present and so the polymerization sites are not blocked by carbon. Increasing the carbon content decreases the film shrinkage, due to the polymerization sites being blocked. As a result, the stress in high-carbon-containing films is lower. It is thought that the crack resistance and gap filling of organic glass films is better than the inorganic because of higher surface mobility of atoms within the polymeric structures. There are fewer O–Si–O crosslinking sites and hence the film is more adapted to withstand a higher level of stress. Addition of carbon to the SOG makes the film more porous and therefore more susceptible to chemical attack as there could be some unreacted silanol groups that react with water.

**Solvents.** One of the most common problems is the wettability of the SOG. This depends on the hydrophobicity of the solvent, which contains 1-propanol and ethanol. Others may also contain 1-butanol and acetone, which increase the solvent hydrophobicity. It is believed that the SOG fills trenches primarily due to the capillary action. In some cases, depending on the solvent type, the capillary action may be counterbalanced or even outweighed by the hydrophobic dewetting forces. To enhance the wettability, the appropriate solvent type may be chosen or the spin time can be increased, but this will be at the expense of planarity and filling.

**Aging.** SOGs generally have a fixed shelf life and need to be stored in cool environments. Age causes the SOG to gel and it becomes very nonhomogenous. The usual problems in using SOG that has exceeded its normal shelf life are striations after coating the wafers and poor film uniformity. Table 2.6 compares the properties of organic and inorganic SOGs.

**Coating.** The coating process depends on the viscosity and the composition of the SOG and the spin speed. The speed can be uniform or ramped from a lower to a higher speed. As the SOG is spun faster the film thickness on flat wafers decreases. On topography wafers a degradation in planarity is observed, especially in narrow metal spaces with increased spin speeds. However, on top of metal lines in the worst case locations, the thickness of the SOG decreases with an increasing spin speed. On the other hand, coating the SOG with a slower spin speed effects the same parameters in reverse. Figure 2.15 shows the effect on the planarization by varying the SOG thickness. This is achieved by spin coating several times. One method of coating

**TABLE 2.6 Comparison of Organic and Inorganic SOGs**

| Property | Organic SOG | Inorganic SOG |
|---|---|---|
| Step coverage | Good | Fair |
| Passivation | Fair | Good |
| Crack resistance | Good | Fair |
| Reliability | Fair | Good |
| $O_2$ plasma resistance | Fair | Good |
| Application | Etchback | Nonetchback |
| Manufacturability | Fair | Fair |

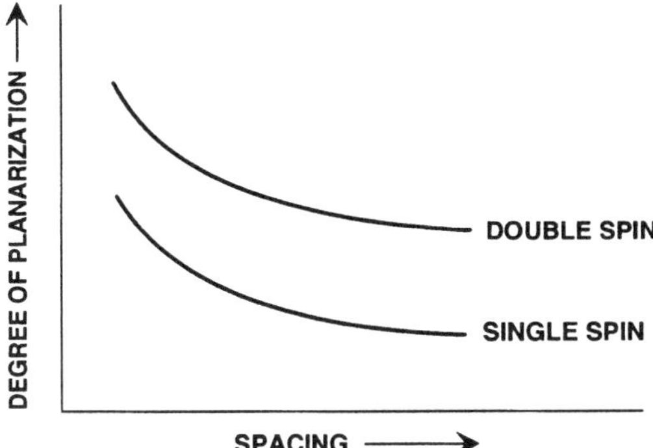


**Figure 2.15** Impact of multiple SOG coatings on degree of planarization.

SOG on wafers is to start with a static dispense followed by ramp and spread cycles. This optimization is needed to ensure that a uniform film of SOG is coated on the wafers. In order to achieve the desired thickness and gap filling, double or triple spin may be required. One of the drawbacks of this procedure is that defects in the film may be generated due to nonuniform wetting of the first layer that might have been cured. In spite of this risk, multiple bakes between spins help in driving solvents out of the SOG. This eliminates the need to bake all the layers of SOG from multiple spins as the thermal treatment may not be enough to completely remove any moisture and solvents.

**Baking and curing.** In general, two reactions[78] take place when the SOG is heated and cured. The first is a hydrolysis reaction where the ethanol is released from the Si-$OC_2H_5$ bonds. The second reaction is a condensation reaction. The thermal treatments the SOG film is exposed to drive the polymerization reaction involving the condensation of silanol groups, resulting in the formation of the Si-O-Si bonds. During the thermal process, the solvents and water outgas from the film. The loss of the solvents and water during the thermal treatment (baking after spin) causes considerable volume shrinkage in the SOG, which creates a high tensile stress in the film. This stress could contribute to crack formation in the glass. Inorganic silicate SOG is typically fragile and cracks or forms gaps in severe topography. Organic SOGs behave differently as they have methyl ($CH_3$) or phenyl ($C_6H_5$) bound to some of the silicon atoms. Presence of the methyl or phenyl groups blocks the polymerization reaction resulting in a more porous and ductile film. Organic SOGs have a higher crack resistance than the inorganic ones. In order to provide sufficient time for outgassing of the solvents and moisture, the postspin bakes and cures must be performed without causing any thermal shock. Baking and curing very quickly could cause the upper layers of the spin on glass to be completely polymerized and prevent adequate moisture and/or solvent evaporation from the bulk of the SOG. Outgassing can lead to trapping of hot volatile compounds beneath a cap of fully polymerized SOG at the upper layer of the SOG. These volatile compounds may then contribute to cracking and other forms of catastrophic stress relief. Therefore, thermal shock at either SOG bake or cure must be reduced. One possible solution is a ramped baking and curing cycle where a gradual increase in temperature. By monitoring the band intensity of the O–H stretch of the SOG using Fourier transform infrared spectroscopy (FTIR),[92] it is possible to quantify the extent to which this condensation reaction approaches completion.

Control of the thermal treatment is very important in general for SOG processing and specifically for a nonetchback option because the SOG is exposed after via etch. Insufficient curing will cause moisture and solvent outgassing through the vias during the metal deposition, which gives rise to a typical defect called "poisoned vias." In such vias, the metal does not completely fill the voids and therefore causes high via resistance and reliability degradation. This suggests that any post-SOG cure temperature steps must be no greater than the SOG cure itself.

## 2.6 Contact/Via Formation

The objective of this process step is to open hundreds of thousands of contacts or vias on a die in order to provide a connection path between the metal interconnects and the transistors. In the processing environment, the contacts/vias need to be opened on varying topography, across the die, within a wafer, and on several wafers in one operational step. Contact process optimization and control are mandatory for ensuring a low defect density, prevention of incomplete etch, and improper contact metallization. With the progression in process and product technology, the contacts are being scaled down and the number of contacts are increasing per unit area. Contact placement, size, metal overlap, and other design rule constraints make the management and control of the contact/via formation step very challenging. The previous steps presented a macroscopic set of problems; the contact and via formation modules present a microscopic set of issues. Localized variations among contacts need to be studied, as these variations differ from contacts/vias that may be separated from each other by only a few microns. The challenge is to ensure consistency and control in definition of contacts and vias using enhanced lithography, etching, cleaning, and contact filling techniques.

In Section 2.1, a brief explanation of the difference between contacts and vias was given. As most of the issues for contacts and vias are similar, the discussion here will focus on contacts. Wherever appropriate, the discussion will also include via-specific issues. In Section 2.6.1, issues pertaining to the contact formation are discussed. In Section 2.6.2, the factors that contribute to contact resistance are examined.

### 2.6.1 Contact/via etch process

Contact and via etch processes[97–99] are critical steps in any integrated circuit process flow. To understand the gamut of variables that affect and modulate the contact/via etch process, a cause and effect diagram, Figure 2.16, is presented that looks at the contact/via formation process as a whole. The contact formation cause and effect diagram highlights some of the salient issues associated with definition of contacts.

The choice of etchers, their configuration, chemistry, and machine settings determine the quality of the etch process and the thoughput time of the etcher. Each of the input/machine parameters, such as chemistry, flow rate, pressure, etc. impact the various output parameters. Output parameters are those that indicate the quality and capability of a etch process. These parameters are etch rate (ER), selectivity (SEL), loading effect (LE), critical dimension loss (CD), profile (PF), uniformity (UNF), defect density (DD), and process-induced charging (PC). Let us examine each of the major machine parameters and examine their impact on the output parameters.

1. *Chemistry.* The gas species used for plasma etching. Has a strong effect on the etch rate, selectivity, profiles, and CD. For etching oxides fluorine (F)–based chemistry is widely used.
2. *Mixture.* The gas ratio determines the polymer formation. For oxide etching, the fluorine to carbon (F/C) ratio is an important parameter.
3. *Flow rate.* The flow rate determines the availability of the reactive species. The most prominent impact is on the loading effect.
4. *Pressure.* The pressure of all the gases in the reaction chamber has a major influence on the profiles of the oxide and the critical dimensions.
5. *Frequency.* Generally the frequency of the RF used is 13.56 MHz. This has an overall effect on the process, but does not modulate any single or multiple parameters strongly.
6. *Power.* Variation in power has marked effect on the etch rates and process induced charging.

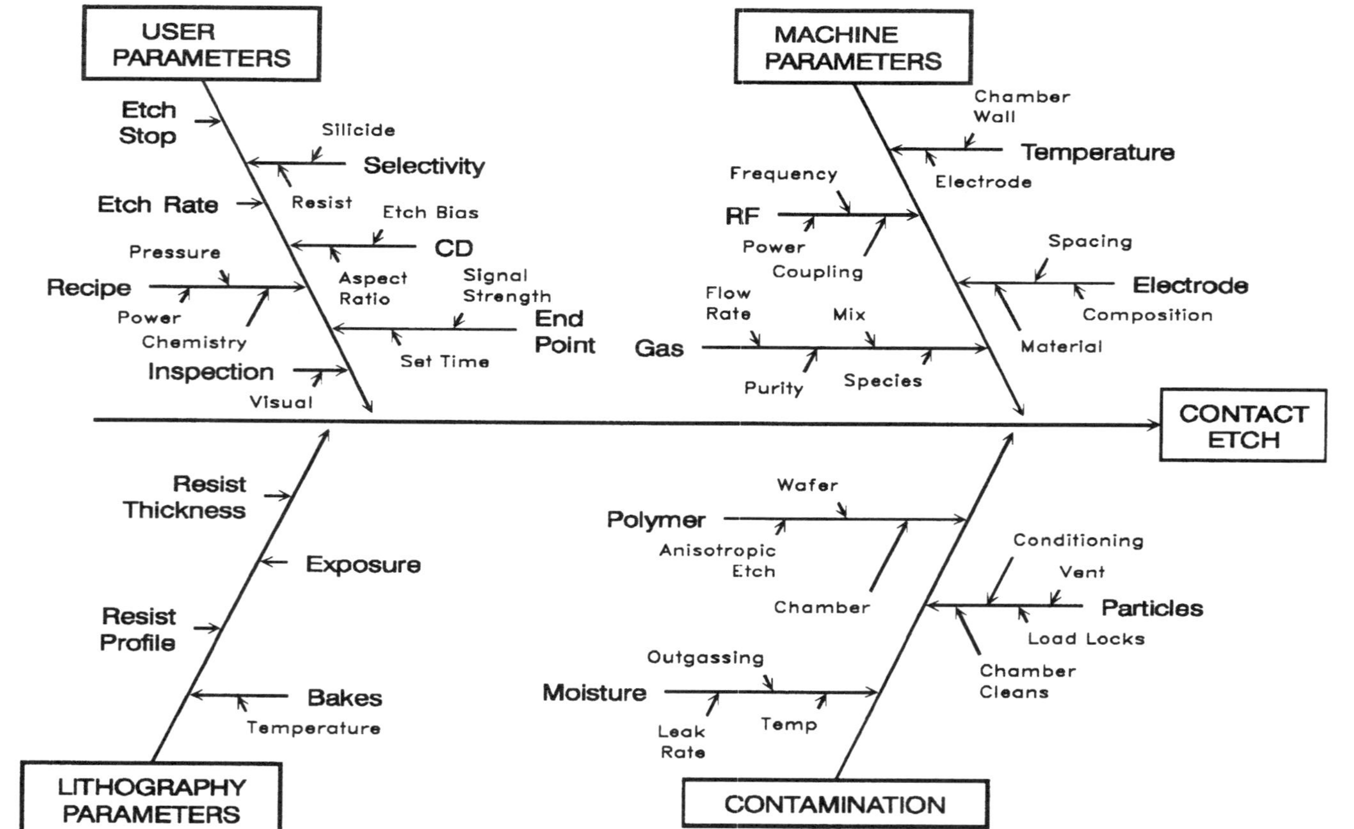


**Figure 2.16** Cause and effect diagram for contact etch process.

**TABLE 2.7 Correlation between Machine and Output Etch Parameters**

| Parameter | Etch rate | SEL | LE | UNF | CD | Profile | DD | Charging |
|---|---|---|---|---|---|---|---|---|
| Chemistry | H | H | M | M | H | H | M | |
| F/C ratio | H | H | M | M | H | H | M | |
| Pressure | M | H | | H | H | H | | M |
| Frequency | | M | | M | M | M | | M |
| Power | H | H | | M | H | M | | H |
| Electrode | M | | | H | M | M | M | |
| Flow rate | M | H | H | | M | M | M | |

7. *Electrode.* The material, type, and spacing of the electrode primarily influences, the uniformity of the etch process across the wafer.

Table 2.7 compares the impact of various machine parameters on the output parameters. H signifies high correlation and M represents medium correlation between input and output parameters. Blank spaces signify weak or no correlation.

**Profile control.** Tapered contacts and vias, to the extent permitted by the device and process sensitivity, have been used for 1 μm or larger process technologies to improve metal step coverage. In submicron technologies, tapered contacts are not a viable option as contact critical dimensions, spacing, metal overlap of contacts, and other design tolerances have been tightened due to device scaling.

How can we achieve near-vertical contact/via profiles? Near vertical contact/via profile control is generally achieved by defining near vertical resist profiles and using a RIE process. In addition, managing polymer buildup during the etch process, through appropriate selection of the F/C ratio,[100] also helps obtain the desired sidewall angles. Polymer coats the sidewall and prevents lateral etch. Removal of polymer after etch may pose a problem. In the case of contacts, acid cleans may be used to strip the polymer, but for vias the choice of cleaning agents decreases as acid cleans would attack the metal layers underneath. Oxygen is added to the fluorocarbon gas in order to improve the etch uniformity, enhance etch rate, and reduce polymer buildup. Reduction in polymer buildup is needed for decreasing the particle density on the wafers and also for minimizing the frequency of the etch chamber cleans. The profiles of the contacts and vias determine the quality and integrity of the contact/via filling process. For example, a blanket CVD W fill deposition process is very sensitive to the contact/via profile and one can expect degraded W step coverage along with a void in bowed or reentrant contacts/vias.

**Differential depths.** The planarization section discussed the advantages of planarization. A side effect of global planarization is the cre-

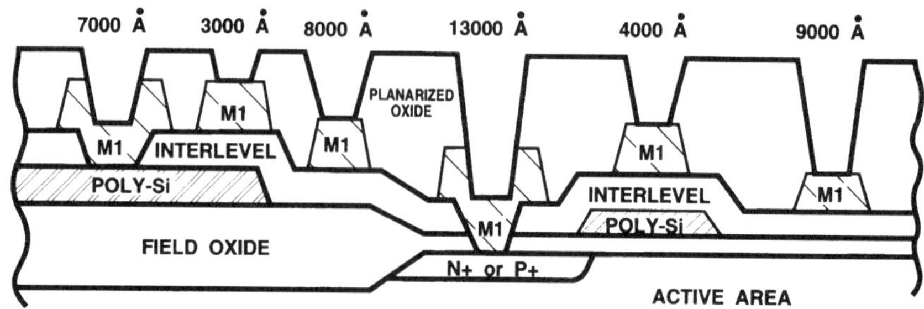


**Figure 2.17** Via depth variation in a multilevel interconnect system.[101] (© *1989 IEEE.*)

ation of contacts and vias with varying depths. Differential depths present a unique contact/via etch scenario. Figure 2.17 shows the via depth differential and the scenario after etch that is a consequence of the planarization.[101] Overetching of the shallow vias can not be prevented as the deep vias take a longer time to be etched. New interconnect materials may be the answer to this problem, along with an effective contact and via clean processes. The clean processes are done prior to metal deposition. During this process, both wet and dry clean processes are used.

**Contact etch stops.** As the implanted source/drain and gate regions offered an unacceptable contact resistance for submicron devices, the use of silicides such as $TiSi_2$ and $CoSi_2$ became necessary. In the contact etch process, for example, the etch must stop on the silicide layer. To prevent significant etch of the silicide in the contacts, especially in the shallow contacts, the selectivity of the silicide film must be very high.[102] If the shallow contacts are overetched so that the salicide layer is also etched, the contact resistance will increase drastically. It has been observed[102] that addition of $H_2$ to the fluorocarbon gas chemistry greatly enhances the selectivity of the $TiSi_2$ film without markedly effecting the oxide etch rate. $SiO_2/TiSi_2$ etch rate ratios of 40:1 have been reported by the formation of a fluorocarbon film over the silicide film. The margin for process control may be small as contact and via etches are still needed of effective endpoint systems. Using endpoint systems for these etches is complicated by the fact that contact/via areas are small.

**Via etch stops.** The vias contain a totally different etch stop. The underlying layer that acts as an etch stop is a metal stack. If aluminum is the stopping layer then it reacts with the fluorocarbon gases to form a Al-F compound. In addition, other aluminum materials may be formed in the vias during the subsequent contact metallization steps that may also increase the via resistance.

### 2.6.2 Contact resistance

Contact formation and its subsequent filling with metal (discussed in Section 2.7, below) significantly modulate the contact resistance. This is monitored either inline or usually at the completion of the process as one of the screens to check the functionality of the finished device. The contact resistance measurement is done on specially designed test structures, containing hundreds of contacts in a chain, and placed in the silicon streets between the dice on the wafer. The variations in contact resistance are used as a flag to diagnose any problems with the contact formation or the contact filling process steps. Control of this parameter is one of the ways to monitor lot to lot performance of the process in general. Noted that there are several other monitors in a process that check the performance of the various steps. The contact resistance test is an integrated monitor and any out-of-control situations would necessitate a closer examination of individual step monitors, at least during the initial stages of failure analysis. To put the whole contact resistance issue in perspective, a cause and effect diagram is given in Figure 2.18.

In an ideal situation, what contributes to contact resistance or via resistance? In the case of a contact it is mainly the sum of the interfacial (interface between contact metal and silicide) resistance and the silicide resistance. In case of a via, it is primarily the interfacial resistance. Assume the entire contact (ideal in formation and with unit dimensions) right from the metal to the silicide layer is a block, with each section having its own resistance, as shown in Figure 2.19. The overall contact resistance is the sum of the interfacial resistance ($R_i$) and the silicide resistance ($R_s$), if we assume that the contribution of the plug ($R_w$) and the metal line ($R_m$) are negligible. In other words, total contact resistance, $R_c$, is given by

$$R_c \approx R_i + R_s \tag{2.12}$$

$R_s$ depends on the silicide formation process and is well controlled. $R_i$ is basically a measure of the contact process health as the significant portion of the contact resistance comes from the interface of the W plug and the silicide. The interface is altered by the etch process, sputter etch, and the W adhesion layer interactions, as explained in Section 2.7.

We can use some generic illustrations to understand the relationship between the contact resistance (applicable to via resistance also) and the following parameters. This list of relationships is not comprehensive but is intended for the reader to appreciate the interactions.

**Aspect ratio.** Figure 2.20 shows the relationship between contact resistance and aspect ratio. The *aspect ratio* of a contact is defined as

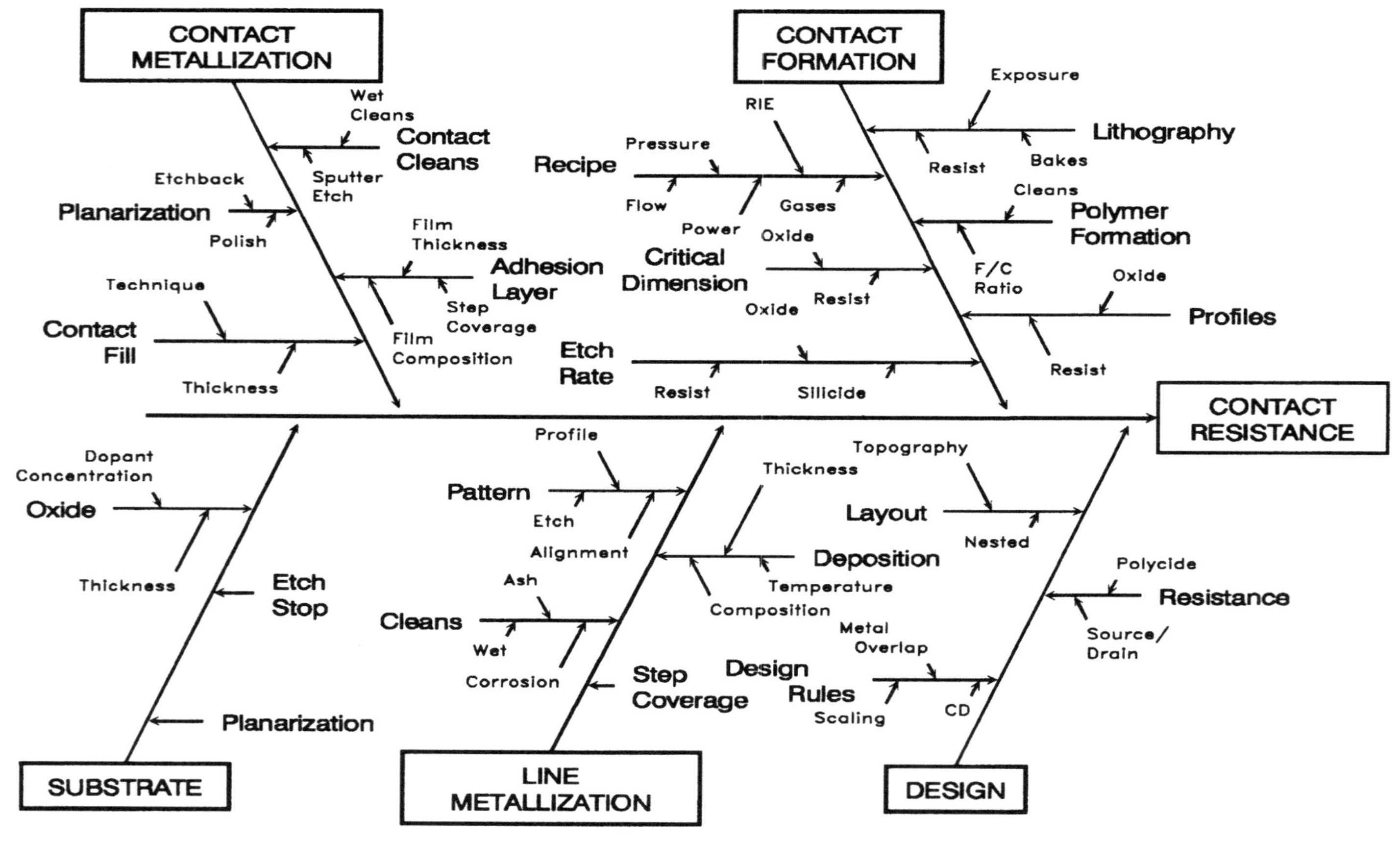


**Figure 2.18** Cause and effect diagram for contact resistance.

| LINE METALLIZATION RESISTANCE | $R_m$ |
|---|---|
| W PLUG RESISTANCE | $R_w$ |
| INTERFACE RESISTANCE | $R_i$ |
| SILICIDE RESISTANCE | $R_s$ |

**Figure 2.19** Contribution to contact resistance by the components in a tungsten-filled contact.

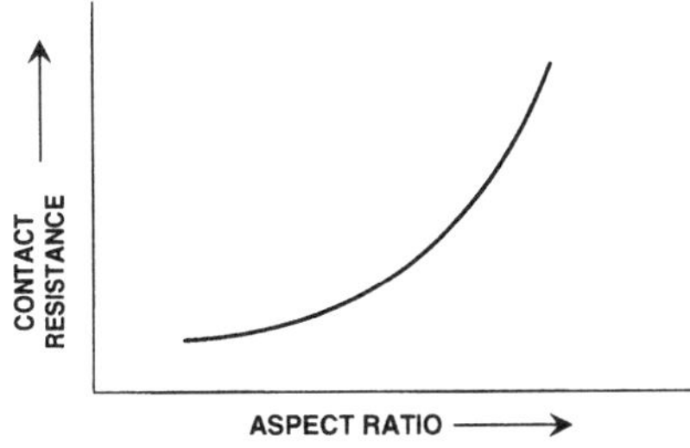


**Figure 2.20** Impact of aspect ratio on contact resistance.

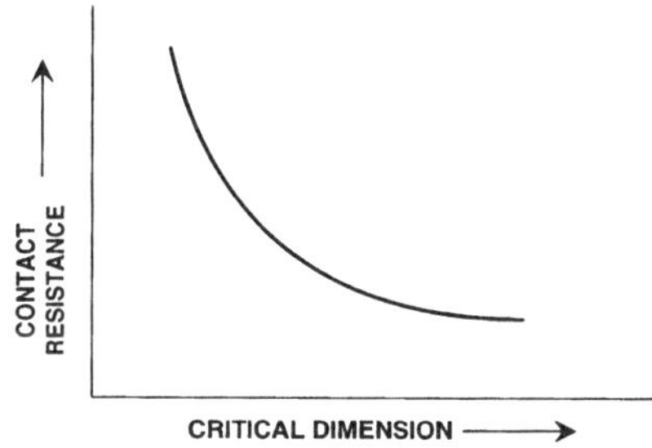


**Figure 2.21** Impact of critical dimension on contact resistance.

the ratio between its depth and width. If the aspect ratio of a contact is high it poses a problem for contact filling. Contact resistance increase may then be dominated by the integrity step coverage of the contact and line metallization.

**Critical dimension.** The contact size, the cross-sectional area at the bottom of the contact, influences the amount of current flow. As the critical dimensions (CD) of the contacts decrease, this cross-sectional area also decreases, offering a higher resistance to the current flow. As in Figure 2.20, this relationship is valid as long as the contacts can be defined in resist, etched, and filled with metal without any loss of contact integrity. The interaction between contact resistance and the critical dimension is shown in Figure 2.21.

**Overetch.** Since contacts have to be etched with differential depths, overetching is essential to ensure that all the contacts are opened and

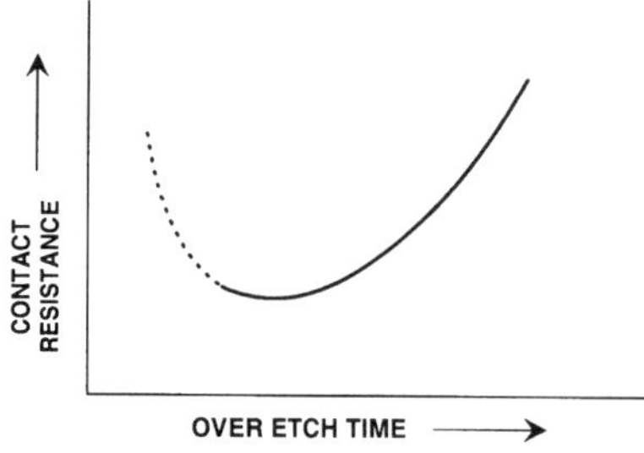


**Figure 2.22** Impact of contact etch time on contact resistance.

therefore the contact resistance decreases. Figure 2.22 shows the contact resistance as a function of the contact overetch time. However, overetch also causes the shallower contacts to be exposed for much greater time to the etch. This does not necessarily create problems as long as the selectivity of the underlying layer is high with respect to the main film that is being etched. As the overetch time increases, the contact resistance also increases as the bottom of the contact is damaged and the etch stop starts to be etched more and more. This relationship is illustrated by the bold curve in the graph in Figure 2.22.

**Sputter etch time.** Just prior to contact metallization, any native oxide is removed by using argon (Ar) sputter etch. The sputter etch process is a physical etch process employing accelerated Ar ions. As the sputter etch time is increased, the contact resistance decreases up to a point. Beyond this, any further increase in sputter etch time merely causes redeposition of material from the sidewalls and increases the defect density. Consequently, the contact resistance starts to increase. Figure 2.23 illustrates the relationship between contact resistance and contact sputter etch time.

This section has explored the issues with the contact formation process and what modulates the contact resistance, which is an important parameter for determining the state of the process. The next section will discuss the contact metallization and interconnect metallization. How these two influence the contact resistance has been shown in the contact resistance cause and effect diagram in Figure 2.16.

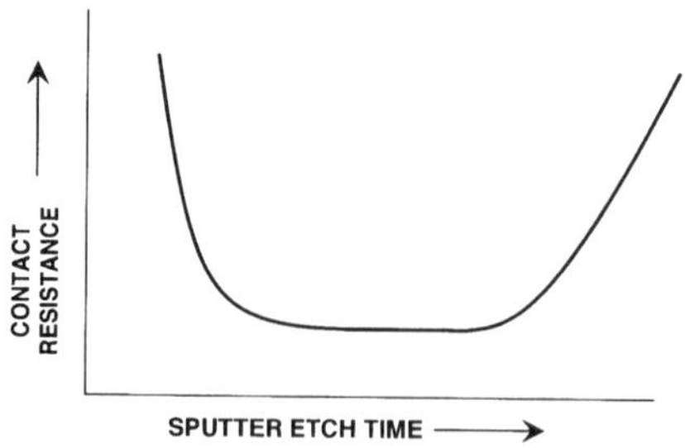


**Figure 2.23** Impact of sputter etch time on contact resistance.

## 2.7 Metals

This last section in the multilevel interconnect process flow (Figure 2.2) deals with contact and interconnect metallization. In terms of process technology development, contact metallization is a recent development and its development was motivated by the migration of device geometries toward submicron dimensions. As this trend continues, more and more development effort will be devoted to understanding the materials and their metallurgical interactions with silicide or other underlying layers in the confines of a contact or a via. In fact, contact metallization has two basic components—contact filling and contact planarization—and each is a subject for extensive research and development. Line or interconnect metallization has been studied for a long time. A wide variety of conductor materials are available for metal interconnects that use aluminum (AlSiCu) alloys and also sandwich structures using refractory and other metals, which are examined in Section 2.7.4, below. For submicron process technologies, the key issue under consideration is the current-carrying capability of the interconnects and other process induced defects. The discussion below is divided into two portions, contact and line metallization.

### 2.7.1 Contact metallization

The design rules for submicron process technologies are rapidly changing that vertical contacts and vias are being presented for metal deposition. In previous technologies, contact and via filling with good metal step coverage was not an issue as these were tapered. As explained above, tapered profiles of the contacts/vias places limits on the maximum packing density. Profile control is difficult to achieve in a manufacturing environment. Vertical contacts and vias cause metal step coverage to be poor if the metal was deposited directly into these holes. In order to improve the step coverage of the metal interconnects, the contact/via holes have to be filled with material that can fill the holes without any voids and can be planarized.[103–106] Metal interconnects running over these plugs will have a superior step coverage and also be able to conduct current through the contact or via.

### 2.7.2 Techniques

There are many possible approaches to contact metallization. The most attractive technologies are the blanket tungsten deposition/etchback and selective tungsten deposition. There are still a lot of manufacturing issues that need to be resolved with these processes. However, until the other approaches are more manufacturable, these two contact metallization methods may continue to be used for the fabrication of submicron devices. The approaches[109] for contact metallization may be categorized as follows.

**Blanket CVD.** Thick conformal deposition of metal can be accomplished by using low pressure CVD techniques. The thick film may be etched back or polished just for contact filling or may be patterned and the interconnect metallization also defined. Such a deposition fills holes and grooves conformally, and the deposition rate is limited by surface reaction and not by mass transport. The main contenders for this method are W and CVD-Al. Although both processes are still in a constant mode of development, blanket W has gained wide acceptance because of its manufacturing ease.

**Bias sputter.** Al and molybdenum (Mo) films have been used to fill contact holes by bias sputtering under high resputtering conditions. One of the drawbacks of bias sputtering is extremely low deposition rate; hence it is not a viable option for a high volume manufacturing process. Barrier layers are necessary to prevent Al and Si interaction. High deposition rates, improved film quality, and contact hole filling over a wide range of aspect ratios would make bias sputter deposition of metals (Al, Mo) a viable alternative to other contact filling techniques.

**Selective deposition.** Selective deposition has an unique property. The film deposition takes place over Si, silicide, and other metals but not over dielectric materials. Selective W deposition has some obvious advantages as explained in greater detail below. However, one of the problems of selective W deposition is filling contacts with varying depths and substrates.

**Other.** Other approaches that have been proposed are the pillar formation, lift-off, blanket CVD polysilicon, and electroless metal deposition. Even though some of these may appear elegant, their use is determined by their manufacturability ease.

### 2.7.3 Contact fill and planarization

The two approaches that will be discussed in some depth are blanket W deposition with etchback and selective W deposition.

**Blanket tungsten deposition and etchback.** As blanket W is widely used to fill submicron contacts and vias in manufacturing processes, let us review the cause and effect diagram for contact filling and planarization using blanket W deposition and etchback as shown in Figure 2.24. W does not adhere to dielectric substrates; therefore an adhesion or glue layer is needed for the W to nucleate on.[19] There are

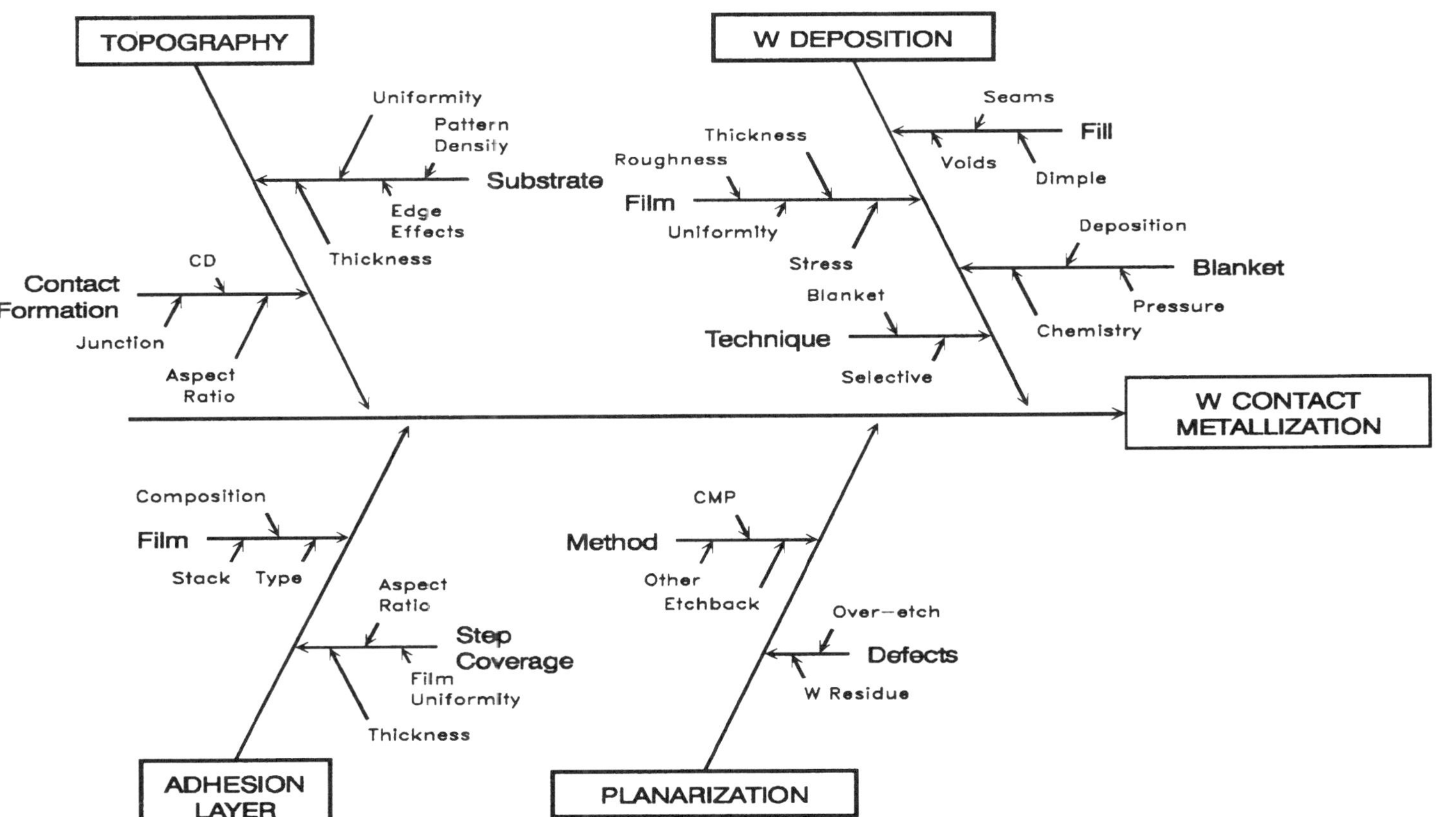


Figure 2.24 Cause and effect for contact planarization using tungsten.

several variations to the adhesion layer. A couple of the commonly used adhesion layers are TiN and TiW. It is this adhesion layer that comes in contact with the silicide source and drain region and contributes to the interfacial contact resistance. The step coverage of the adhesion layer in high-aspect ratio contacts and vias is of concern as poor step coverage may cause W-related defects if it reacts with underlying layers. This issue may be overcome by combining conventional glue layers with CVD-$WSi_x$ or other thin films.

In deposition process,[107] W is deposited in a low-pressure, low-temperature CVD system using tungsten hexafluoride ($WF_6$). The chemical reaction involves the reduction of $WF_6$ using both $H_2$ and $SiH_4$ in a two-step process to form W. The deposition process has to be carried out in two steps employing both the properties of silane and hydrogen reduction chemistries. The majority of the film is deposited by the hydrogen reduction of $WF_6$ after a few hundred angstroms are first deposited by silane chemistry. The two chemical reactions proceed as follows for the silane and hydrogen, respectively:

$$2WF_6 + 3SiH_4 \rightarrow 2W + 3SiF_4 + 6H_2 \tag{2.12}$$

$$WF_6 + 3H_2 \rightarrow W + 6HF \tag{2.13}$$

Table 2.8 puts the benefits of each of these reactions in perspective. The first step in the blanket deposition process is the deposition of a very thin W film. This film is a few hundred angstroms thick and the reaction proceeds using silane, as shown in Equation (2.12). After this step, the bulk of the W film is deposited by hydrogen reduction of $WF_6$. The W must be deposited without voids in the contacts.

One of the drawbacks of the blanket deposition is the formation of dimples in the center of the plugs. This dimple formation is not only dependent on the deposition process, but also on the contact geometry. The W film is rough. Roughness of the film impacts the dimple formation in the W plugs. The rougher films seem to have a smaller dimple size as compared to smoother films.

In the case of via filling, blanket tungsten is again used. However, if the tungsten via plugs are formed on Al interconnects, the control of via resistance is more difficult than if the plugs were to be defined over W interconnects. During the adhesion layer deposition and the

**TABLE 2.8 Comparison of W Deposition Processes**

| Parameter | Silane | Hydrogen |
|---|---|---|
| Deposition rate | High | Low |
| Step coverage | Poor | Good |
| Nucleation | Glue layer | W |
| Thickness | Very thin | Thick |

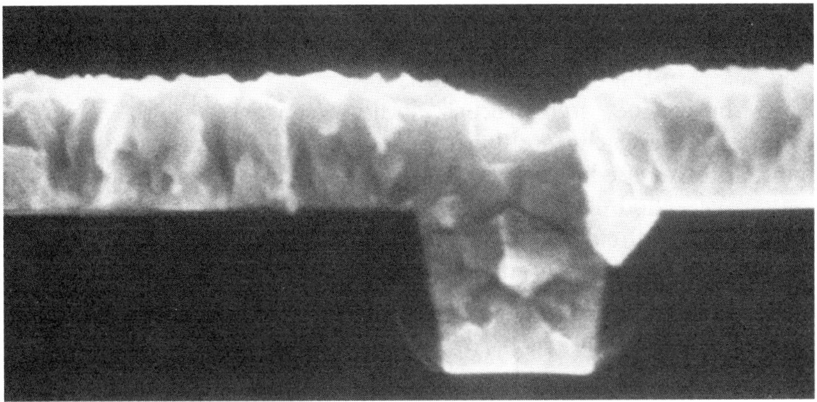

**Figure 2.25** Cross section of a tungsten film in a contact.

subsequent W deposition process, nonvolatile intermediate species, e.g., aluminum halides, form at the bottom of the via on the Al surface. The presence of these compounds may increase the via resistance. Figure 2.25 shows a scanning electron micrograph of the W film after deposition in a contact.

As the W is deposited all over the wafer, it must be etched back, leaving W only in contact and via holes. Several techniques[19] have been demonstrated to perform this planarization, like chemical mechanical polish.[114] Each of these has its own technical and manufacturing problems. Tungsten etchback is still widely used for contact planarization.

The simplest way to etchback is to etch the W directly, without using any sacrificial layer. The process does appear to be simple in concept, but has several issues that may affect the contact/via resistance and metal step coverage causing yield degradation.

**Bulk film etch.** The incoming W film thickness is dependent on the smallest and the largest aspect ratio contacts. W must fill these contacts without any voids and have sufficient thickness for the etchback process. As the contact holes become narrower, the W deposition process becomes difficult as voids may form. Wider contacts, on the other hand, need a greater amount of material to completely fill the contacts. The W etchback process is performed using fluorine and chlorine (for example, $SF_6$, $Cl_2$, and He) containing gases. The purpose is to leave a planar plug in the contact or via with no residue.[111] Usually the etchback process is done in multiple steps. The first step is aimed at etching the bulk of the W film in an isotropic manner with a high etch rate. To achieve an isotropic etch, fluorine-based gases are employed. The isotropic nature of this step also smooths the inherent roughness of the W film. The second step is an anisotropic etch employing chlorine-containing gases with a high selectivity to the adhesion layer. Further discussion on the adhesion layer etch is presented below in the Overetching section.

As in the contact etch process, shallow contacts may be severely overetched during the etchback processes.[19] This may cause catastrophic failures of shallow contacts as the W and the adhesion layer get severely etched from the plug. As the adhesion layer is etched out, the underlying silicide layer is exposed to the etch chemistry and is removed. Removal of the silicide layer causes the contact to have a high resistance and therefore fail. The polishing of W would overcome such problems.

**Microloading.** This is a problem quite specific to the etchback process. Localized variations in etch rates can exist and cause areas of complete removal of plugs while at the same time there may be areas where the W is still not fully etched back. Microloading occurs during the final clearing of the W over the dielectric substrate. Microloading is a localized effect where the etch rate differences occur due to localized consumption or depletion of reactant neutrals and appears to be independent of the ion-assisted surface reaction processes.

**Overetching.** Over etching is essential to ensure all the W residues are completely removed. Residues left on the wafer may cause interlevel shorts and other defects that may nucleate on these sites. During the overetch step, the selectivity of W to the adhesion layer plays an important part. The amount of adhesion layer etchout depends not only on the selectivity but also on the thickness of the adhesion layer itself. The nonuniformity of the overall W deposition and etchback process will impact the contact/via resistance variation and even the step coverage of the metal interconnects. The schematic diagram shown in Figure 2.26 illustrates a possible scenario if the adhesion layer were to be overetched that could lead to reliability degradation.[19,128] The etchback process may be combined with the def-

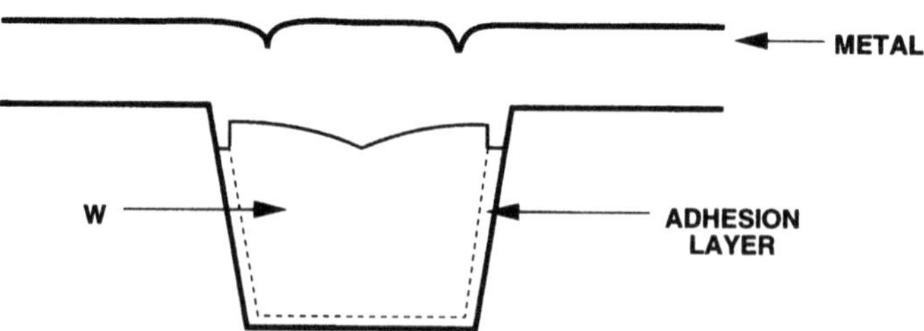


**Figure 2.26** Impact on metal step coverage due to blanket tungsten deposition and etchback process variations.

inition of W interconnects, instead of Al. W interconnects, along with the rest of the materials that may be used as interconnects are discussed in Section 2.7.4, below.

**Selective W deposition.** This method [115–117] of filling contacts and vias has been investigated by many researchers and it should become a viable process once some of the manufacturing issues are resolved. Selective deposition allows W to form on silicon, silicide, or metal areas in the bottom of contacts or vias and not over dielectric films. The W is formed by hydrogen reduction of $WF_6$, as given by Equation (2.14), only after an initial reaction of $WF_6$ with Si in the contacts. The chemical equation for this reaction is given as follows:

$$2WF_6 + 3Si \rightarrow 2W + 3SiF_4 \tag{2.14}$$

Prior to the deposition of W, any silicon dioxide is removed from the bottom of the contacts and then the reaction is allowed to proceed. During the initial few minutes, the W deposition occurs by the Si reduction of $WF_6$, as shown above. This is a self-limiting step; after a few tens of nanometers of W are formed, the reaction is stopped as the W acts as a diffusion barrier and prevents the Si from reaction with the $WF_6$. After this initial phase, the remaining amount of W required is formed by the hydrogen reaction, as shown in Equation (2.14).

Improvement of selectivity is necessary to ensure that only the Si, silicide, or metal layers are deposited with W and not $SiO_2$. The selective feature of W deposition depends on numerous parameters. Degradation of this feature is observed when $WF_6$ or $H_2$ partial pressure, deposition time, deposition temperature, and the ratio of the Si to $SiO_2$ area are increased. This suggests that the management of the entire deposition is crucial to ensure that the selective property is maintained.

Besides improvement in deposition, another technique[113] is preventing W formation on the dielectric oxides. This is accomplished in two ways. First, change the surface of the oxide just prior to W deposition. The oxide surface may be altered by a HCl plasma or methanol pretreatment.[113] This has the effect of controlling the dielectric surface states that effect the selective deposition property. Second, cover the oxide with some other material that suppresses the W nucleation.

Cleaning of the contact/via surfaces is needed to improve the contact/via resistance as any residual contamination or native oxide present on these surfaces may prevent the formation of W. Recent work[112] suggests that the use of dimethylhydrazine (DMH) gas in the CVD W deposition process can remove the native oxide on the Si surface. DMH cleaning showed a drastic reduction in the sheet resistivity of the W layer and had some positive effects on the junction leakage.

### 2.7.4 Line metallization

This is the last step in the sequence of operations for multilevel interconnect formation. After all the interconnect layers are defined, a passivation film is deposited and the device is tested for functionality. After testing the devices proceed to the assembly sites for packaging. Interconnects serve the following key functions:

1. Provide paths to contact junctions and gates
2. Interconnection between device cells
3. Input and output pads for connection to external signal sources via the package leads

To satisfy these functions, interconnects should have a low resistivity for faster signal propagation, be reliable under various operating conditions, and adhere well to Si and $SiO_2$. Moreover, they should be compatible with the other fabrication processes.

**Aluminum interconnect problems.** As mentioned above, several materials are being used for the interconnects. The question is, Can we continue to use Al interconnect schemes for submicron devices? Several technologies have begun to use W and even Copper (Cu) interconnects as an alternative to aluminum because of the problems associated with Al interconnects.[126,127]

Aluminum silicon alloys have become the primary materials for interconnects because pure aluminum has great affinity for silicon and therefore consumes it from the junctions. To prevent this, Al has been saturated with Si. Even though the problem of junction spiking was solved with the use of an Al-Si alloy, the problem of Si precipitation still exists in contacts that do not use barrier metallization. These precipitates have a detrimental effect on the contact resistance and also on the electromigration performance of interconnects. For enhancing the electromigration performance of the interconnects, Cu is added to Al. The amount of Cu that can effectively prevent electromigration degradation is typically around 0.5 percent. Any higher percentages of Cu cause difficulties in dry etching, corrosion, and ultimately reliability degradation. A more detailed explanation of the impact of Cu on the electromigration performance of Al interconnects is presented in Chapter 4 of this book. Besides electromigration issues, Al interconnects have suffered from their susceptibility to form hillocks (stress relief mechanism) and voids.

In spite of all the problems with Al interconnects, current metallization schemes have attempted to make them more robust by adding shunt layers like Ti and barrier layers like TiW or TiN. For lithography improvements, an antireflective layer is added to the stack. The current submicron processes are not just two-level or three-level metal systems anymore. In fact, on closer examination, the metal stack often consists of several metal layers. Therefore several metal CVD or sputter machines must be used for the deposition of the metal layers starting all the way back to the silicide process step.

**Cause and effect analysis.** There are several interrelated process steps that can influence the quality of the interconnect definition. The cause and effect diagram, shown in Figure 2.27, puts these in perspective. The issues associated with wiring layout, scaling, etc. were discussed at length in Chapter 1. The reliability and manufacturing issues will be dealt with in other chapters. It is the time delay for signals to propagate through the interconnects that increasingly determines the performance of the device and not so much on the switching speeds of the transistors. Therefore, metallization becomes a very key component of VLSI/ULSI technology because it impacts the device density and performance.

Let us discuss aluminum interconnect definition by using the cause and effect diagram that is given in Figure 2.27. This diagram may have to be modified should other materials be used for interconnects. But the basic approach will still be the same.

Substrate and contact metallization issues were discussed in earlier sections with more detailed cause and effect diagrams and in-depth analysis of the issues. Therefore, let us focus on the two important items of interconnect definition, metal deposition and metal etch/cleans.

**Metal deposition.** The quality and integrity of metal deposition is extremely critical for ensuring high yielding and reliable parts. Another cause and effect diagram, shown in Figure 2.28, is used to explain the integrity of sputtered metal. Metal is deposited on wafers by sputtering Al and other elements from a common target material. In the sputter chamber, Ar ions (used for sputtering) dislodge the Al and other elements from the target and these are deposited over the wafer. The appropriate target material with the desired composition is selected since it is the source material for the metal deposition. It erodes with time as sputtering progresses. Sputtering produces films with an excellent compositional uniformity across the thickness of the film.

Heating the substrate improves the step coverage because the surface mobility of the deposited metal atoms is proportional to the temperature. This helps in improving the step coverage, but only to a certain limit. The importance of step coverage in thin films has been discussed above. In the case of Al alloys, step coverage of the film is strongly dependent on the temperature the wafer is exposed to during

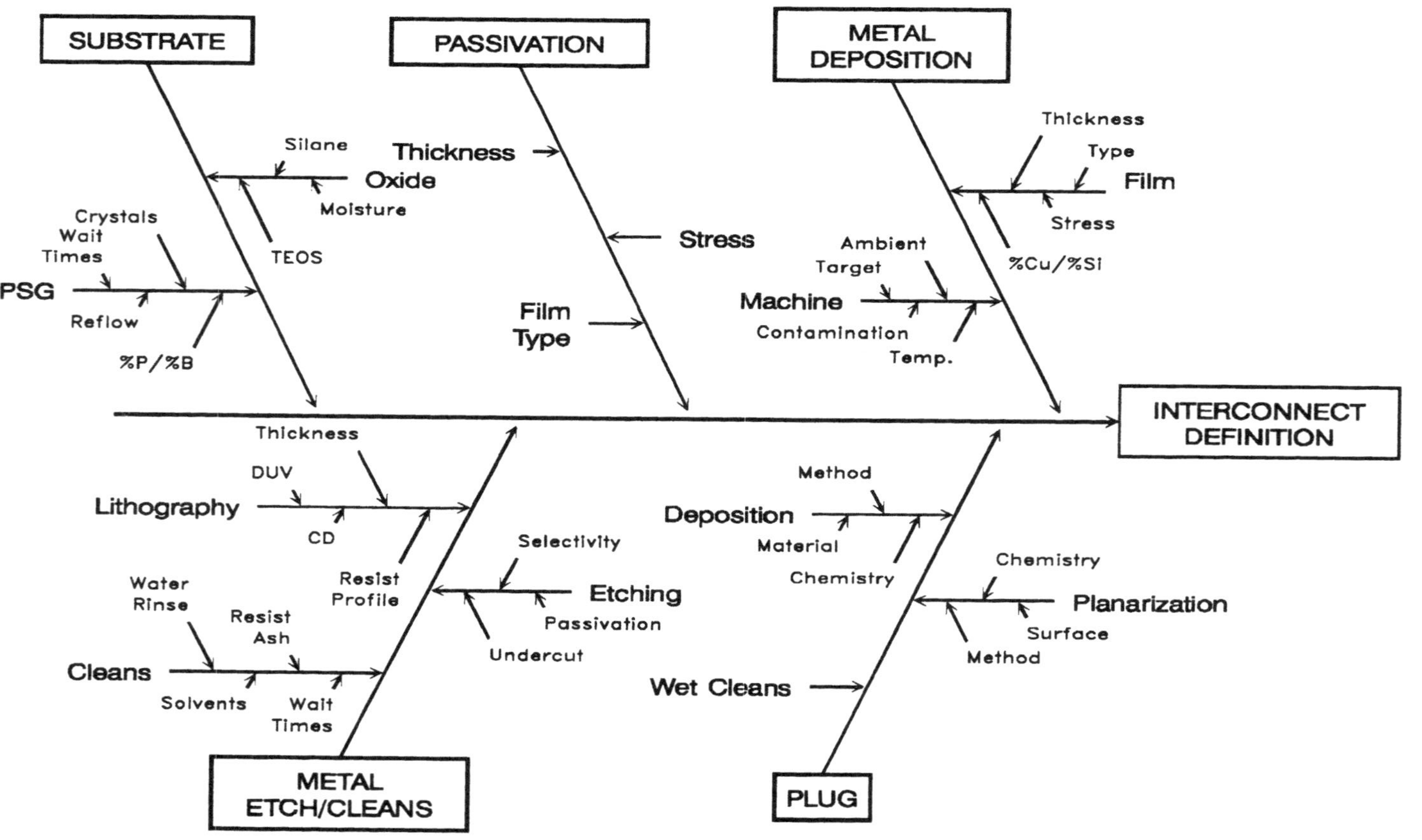


**Figure 2.27** Cause and effect diagram for interconnect definition.

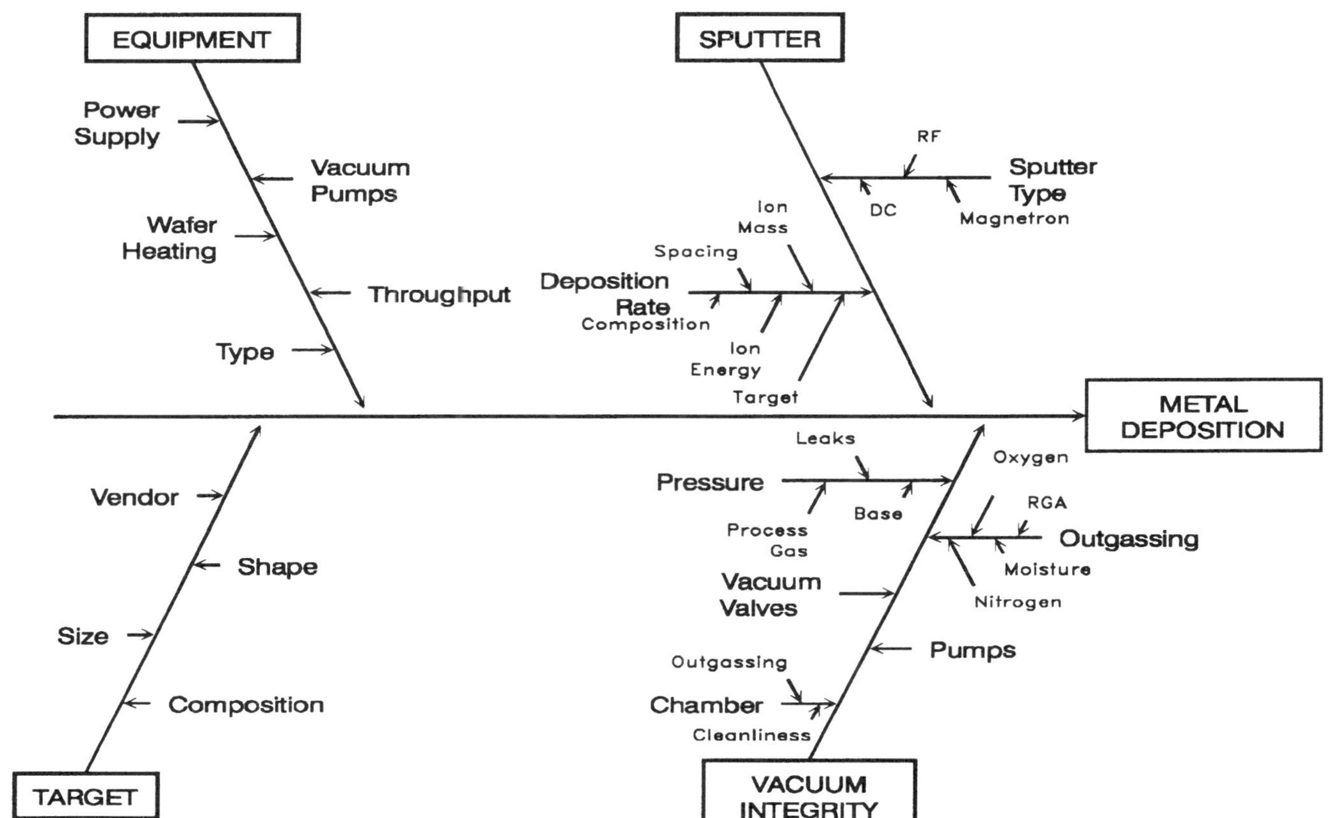


**Figure 2.28** Cause and effect diagram for metal deposition.

deposition. As temperature increases, surface mobility of the deposited metal atoms increases and therefore the step coverage is enhanced. However, increased deposition rates may degrade the step coverage because surface mobility rates of atoms on the wafer surface are decreased due to the arrival of more and more atoms. Increased heat causes hillocks to form in aluminum films due to stress relief mechanisms. Another property that is changed by temperature is the grain size of the metal. The grain size increases with an increase in temperature. The grain size variation is not just dependent on temperature but also on the surface topography, sputter deposition conditions, and vacuum integrity.

The sputtering process is complicated since a lot of variations in the sputtering parameters may produce films of inferior quality. For example, the sputtering process depends on the equipment used, and the equipment parameters as shown in Figure 2.28. The film properties depend on the integrity of the vacuum system used. The base pressure of any vacuum system indicates the level of system cleanliness and the leak rate in the system chamber. During the deposition of the metal, any leaks or chamber outgassing will alter the film properties. The residual gas contribution to the sputtering gas may result in the incorporation of nitrogen and oxygen in the metal films. Refractory metals, aluminum, and its alloys are readily oxidized and the resulting film resistivities and stress may be high. Residual gas contamination may cause a decrease in film reflectivity also. A residual gas analyzer (RGA) may be used to monitor the chamber condition prior to deposition.

**Metal etch.** By far the most crucial aspect of interconnect definition is the process of etching[119–121] Al alloys with shrinking pitches and tight critical dimension control. The etching of Al alloys is very important and difficult as poor management of pre-etch, etch, and postetch treatments could cause metal corrosion and other problems, specifically in Al alloy interconnects. Corrosion is a major problem that affects metal interconnect integrity and has been a subject for extensive study.[118] Hence, let us examine the issues associated with metal etch from a corrosion perspective. Further discussion on corrosion is given in Chapter 3 of this book. The whole sequence of interconnect definitions may be divided into three parts, that is, premetal etch, bulk etch, and postmetal etch.

**Premetal etch.** The pre-etch processing deals with the resist processing. The resist type (e.g., vendor, single, or multiple layers) impacts the etch process and the output parameters, such as line widths, etc. This becomes more of an issue with the definition of each interconnect level. The metal thickness increases depending on the interconnect level and usage. With increased thickness comes a demand for increased resist thickness, tight critical dimension (with decreasing metal pitches), and profile control. All this has to be done on a highly reflective surface (even with the use of an antireflective layer) which may not be perfectly planar (depends on the planarization process). Increased metal thickness translates to increased etch time and a chance for resist profile degradation during the etch. The resist has to harden after exposure and development for it to withstand the etch process. Unhardened resist will reticulate during the etch. Hardening is accomplished in a deep ultraviolet (DUV) system where the temperature and exposure to UV light forms a crust on the resist. This crust protects the bulk of the resist and prevents reticulation.

**Bulk etch.** To pattern Al and its alloys, chlorine (Cl)–containing plasmas are used in a RIE mode. This form of etching has been quite successful in giving good metal profiles and tight critical dimensions—both of which are key requirements for increased packing density of interconnects. But using Cl chemistry to etch aluminum is not without major problems. Al and Al alloys patterned in such a manner have been found to corrode very rapidly on exposure to the atmosphere. The etching is further complicated with the use of different materials to strengthen the interconnect capability of the aluminum, in the form of barrier, shunt, and antireflective layers.

The corrosion problem is aggravated with the use of Al-Cu alloy. The addition of Cu in the aluminum is necessary for reducing the formation of hillocks/voids and for improving the overall electromigration of the interconnect. Aluminum oxidizes readily to form a thin layer of native oxide over it. This must be removed and is done with use of $BCl_3$. Once the native oxide is removed then the Cl plays a major role in etching the aluminum. In order to obtain an anisotropic etch that results in near-vertical sidewall profiles, two techniques are used. First, a polymer film is continually formed on the wafer surface during the etch. This film is removed in flat regions by ion bombardment to allow the etching of Al, but remains on sidewalls to prevent undercut. The etch chemistry and conditions control the polymer formation. The polymer on the top of the metal needs to be removed so that the etching may proceed. This is accomplished by using nitrogen ions to bombard the horizontal surfaces, thereby clearing the polymer films. Since the clearing of the polymer takes place normal to the incident beam of nitrogen ions, the profile of the metal pattern is strongly dependent on the angle of the topographic step.The metal etch contains an overetch component that is needed to clear any residual metal from the surface. Otherwise, these residual metal filaments may cause shorts and make the devices fail.

**Postmetal etch.** The postetch process determines the success or failure of interconnect definition. To prevent the wafers from corroding on exposure to the atmosphere, the wafers are passivated prior to being removed from the etcher. This passivation is done by substituting the Cl absorbed in the resist with F from the passivation gas. After this substitution is done, the wafers are removed from the etcher and rinsed in deionized water to remove all the excess Cl. Immediately following this, the resist is stripped using both plasma and organic resist strippers.

The management and control of corrosion has taken the lead role in the Al-based interconnect definition. As new materials are developed to substitute for the Al-based interconnects, new etching techniques will also evolve. As with any other process, the issues pertaining to process control will play a paramount part in the interconnect definition. Corrosion that manifests itself much later in time after the devices have been shipped can be a nightmare for both suppliers and customers.

**Metallization applications.** There are several applications (Refs. 108–110, 122, 123) where metal films are needed. Table 2.9 shows some of the common metals, their compounds, and their applications. It is being presented here for illustration purposes only. These materials may be used for other applications, and an exhaustive list is beyond the scope of this book. Metallization schemes (the stack composition) may vary depending on the device reliability requirements. In a generic situation, this table may represent a cross-sectional image of the entire metallization stack in a multilevel interconnect system. By choosing a stack approach to interconnect metallization, the reliability of the interconnect is enhanced. Only Al or its alloys by themselves suffer from problems that are detrimental to the reliability of the entire device.

The issues with silicide and contact metallization have already been discussed along with the barrier/adhesion layer issues pertaining to the W contact filling and planarization. Let us examine the metal interconnect stack, specifically the contributions of the barrier layer, primary conductors, shunt layer, and the antireflective layers.

**TABLE 2.9 Metallization Applications**

| Application | Material |
|---|---|
| Silicide/gate | Ti, Co, Ni, W, Ta, Mo |
| Contact fill | Al alloy, W, Si, $WSi_2$ |
| Barrier/adhesion | TiN, TiW, $WSi_x$, TaN |
| Shunt | Ti, W, TiW |
| Primary conductor | Al alloy, W, Cu, Au |
| Antireflective | TiN, TiW |

**Barrier metal.** Barrier metal layers are used to prevent an interaction and reaction between two adjacent materials. If no barrier layers are present, the reaction takes place in the form of diffusion of one of the metals into the other to form a new material. In analysis of the barrier layers, Nicolet[124] proposed a classification of the barrier layers. These films may be separated into three groups:

1. *Passive barriers.* These are chemically inert against the adjacent layers.
2. *Sacrificial barriers.* These materials take part in the reaction and are consumed in it.
3. *Stuffing barriers.* These minimize the diffusion along the grain boundaries.

The requirements for barrier metal layers are as follows:

1. Good adherence to Si, silicide, oxide, and other metal layers.
2. Low contact resistivity is very important since this layer makes contact to the Si or silicide layer.
3. The barrier metal films must be able to withstand thermal processing, prevent interdiffusion, and also be easy to etch.

TiN is widely used as a barrier and as an adhesion layer in processes that use W for contact filling. The advantages of combining these are essentially simplicity in processing. TiN is deposited either by reactive ion sputtering or nitridation of sputter-deposited Ti in a nitrogen ambient. The interactions of the barrier layers with aluminum show that TiN has the capacity to absorb Al and react with it, forming an intermetallic compound. This absorption of Al increases with temperature and therefore prevents the diffusion of Al across the barrier.

**Primary conductor/shunt layer.** Aluminum and its alloys—AlSi, AlSiCu, and AlCu—are extensively used as interconnect materials, primarily because of their high conductivity, ability to form low resistance ohmic contacts to both $p^+$ and $n^+$ contacts, and good adhesion to silicon and silicon oxide. The electromigration performance of the Al interconnect is enhanced by the addition of Cu to Al. As the devices are scaled to submicron regimes, the interconnections may occupy a significant area of the chip and hence the reliability of these is of paramount importance. The requirements, therefore, are changing for submicron interconnects. These need to possess the advantages of Al interconnects, but must be more electromigration-resistant, must be easy to etch, and must overcome the corrosion problems after etch that affect Al interconnects. Several interconnect schemes are in use to make the interconnects more robust. Some metallization schemes[127] use W to cover the Al interconnects to prevent hillock formation during thermal stress. Titanium (Ti) layers that are harder and more electromigration-resistant than Al may also be used. Therefore, if the Al interconnects suffer from defects that present a current restricting path, the Ti or such layers may still conduct the bulk of the current and prevent the device from failing functionally.

Elegant interconnect schemes not only must make the interconnect system more reliable, but also be easy to etch and overcome the problems associated with the etch. Al and its alloys are etched in a chlorine-containing chemistry that causes corrosion, if the postetch treatments are not managed with care.

**Antireflective layer.** If the Al surface is highly reflective and coupled to varying and uneven topography, the reflection of UV light will be uncontrolled and cause CD loss due to unwanted exposure. To prevent this, an antireflective layer is deposited over the metal. This reduces the reflectivity of the surface and helps make the lithography process more controllable.

## 2.8 Summary

Widespread use of multilevel interconnect technology for enhancing the performance of advanced integrated circuits has hastened process development efforts in several areas. New dielectric and interconnect materials in combination with plasma etching and planarization methods have enabled the successful fabrication of submicron devices. The challenge is to develop and transfer to high-volume manufacturing process technologies suitable for sub-half-micron geometries.

## References


1. D. B. Davis, "Technologies ride a fast track into the 1990s," *Electronic Business,* December 1989.
2. J. M. Martinez-Duart and J. M. Albella, "Metallization technologies for ULSI," *Vacuum,* Vol. 39, No. 7/8, 1989.
3. P. Geraghty and W. R. Harshbarger, "Metal interconnects for ULSI integrated circuits," *Proc. Tech. Prog. Nat. Elect. Packg. Prod. Conf.,* NEPCON West, 1990.
4. S. R. Wilson et al., "Multilevel interconnects for integrated circuits with submicron design rules," *SPIE Advanced Processing of Semiconductor Devices II,* Vol. 945, 1988.
5. K. Mitsuhashi et al., "Interconnection technology for three-dimensional integration," *Japanese J. App. Phy.,* Vol. 28, No. 4, April 1989.
6. M. B. Small and D. J. Pearson, "On-chip wiring for VLSI: Status and direction," *IBM J. Res. Develop.,* Vol. 34, No. 6, November 1990.
7. D. J. Bartelink and K. Y. Chiu, "Interconnects for submicron ASIC's," *Intl. Symp. on VLSI Tech. Sys. and Appl.,* 1989.
8. K. Osinski et al., "A 1 μm CMOS process for logic applications," *Philips J. Res.,* Vol. 44, 257–293, 1989.

9. R. E. Oakley, "Advanced interconnect for VLSI," *ESPRIT '84: Status Report of Ongoing Work,* 1985.
10. P. L. Pai and C. H. Ting, "A framework for multilevel interconnection technology," *IEEE VMIC,* 1988.
11. A. N. Saxena, "Update and future directions for VLSI multilevel interconnection technology," *State-of-the-Art Seminar,* Course Coordinator: T. E. Wade, 1988.
12. H. Kotani, "Metallization for ULSI," *20th Conf. on Solid State Devices and Materials,* 1988.
13. G. De Santi, "Interconnections technologies for VLSI circuits," *ETT,* Vol. 1, No. 2, March–April 1990.
14. J. C. Sum et al., "A two micron pitch double-level metallization process," in L. B. Rothman (ed.), *Proc. Symp. Multilevel Metallization, Interconnection and Contact Technologies,* ECS, Vol. 87-4, 1987.
15. P. Cagnoni et al., "A triple metal interconnection process for CMOS technology," *VMIC,* June 1989.
16. D. Fisher et al., "A submicron CMOS triple level metal technology for ASIC applications," *IEEE Custom Integrated Circuits Conf.,* 1989.
17. T. Nishida et al., "Multilevel interconnection for half-micron ULSI's," *VMIC,* June 1989.
18. A. N. Saxena and D. Pramanik, "VLSI multilevel metallization," *Solid State Technology,* December 1984.
19. S. Wolf, *"Silicon Processing for the VLSI Era—Volume 2: Process Integration,"* Lattice Press, 1990.
20. H. P. W. Hey, "Progress in plasma CVD technology," *Microelectronic Manufacturing and Testing,* January/February 1990.
21. B. Gorowitz et al., "Recent trends in LPCVD and PECVD," *Solid State Technology,* October 1987.
22. P. H. Singer, "Depositing high quality dielectrics," *Semiconductor International,* July 1989.
23. Peter Burggraaf, "Thin film technology for advanced semiconductors, part 2b: Dielectrics," *Semiconductor International,* July 1985.
24. J. M. Blum, "Chemical vapor deposition of dielectrics: A review," *Proc. of the 10th Intl. Conf. on CVD,* 1987.
25. D. E. Ibbotson et al., "Oxide deposition by PECVD," *SPIE, Monitoring and Control of Plasma-Enhanced Processing of Semiconductors,* Vol. 1037, 1988.
26. G. B. Raupp and T. S. Cale, "Step coverage in plasma enhanced deposition of silicon dioxide from TEOS," *VMIC,* June 1989.
27. N. Lifshitz and G. Smolinsky, "Detection of water-related charge in electronic dielectrics," *Appl. Phys. Lett.,* Vol. 55, No. 4, July 1989.
28. H. P. W. Hey et al., "Ion bombardment: A determining factor in plasma CVD," *Solid State Technology,* April 1990.
29. S. U. Kim and Albert F. Puttlitz, "Plasma process induced device degradation," *IEEE Transactions on Components, Hybrids and Manufacturing Technology,* Vol. CHMT-8, No. 4, December 1985.
30. D. Pramanik, "CVD dielectric films for VLSI," *Semiconductor International,* June 1988.
31. K. H. Hurley et al., "BPSG films deposited by APCVD," *Semiconductor International,* October 1987.
32. R. G. M. Penning de Vries and K. Osinski, "Enhanced process window for BPSG flow in a salicide process using a LPCVD nitride cap layer," *ESSDERC,* 1989.
33. F. S. Becker et al., "Process and film characterization of low pressure tetraethylorthosilicate borophosphosilicate glass," *J. Vac. Sci. Technol.,* B4 (3), May/June 1986.
34. K. H. Hurley, "Process conditions affecting boron and phosphorous in borophosphosilicate glass as measured by FTIR spectroscopy," *Solid State Technology,* March 1987.
35. W. Kern and R. K. Smeltzer, "Borophosphosilicate glasses for integrated circuits," *Solid State Technology,* June 1985.

36. W. Kern and W. A. Kurylo, "Optimized chemical vapor deposition of borophosphosilicate glass films," *RCA Review,* Vol. 46, June 1985.
37. R. Burmester et al., "Reduction of titanium silicide degradation during borophosphosilicate glass reflow," *ESSDERC,* 1989.
38. P. B. Johnson et al., "Using BPSG as an interlayer dielectric," *Semiconductor International,* October 1987.
39. J. S. Mercier, "Rapid flow of doped glasses for VLSIC fabrication," *Solid State Technology,* July 1987.
40. W. Kern and G. L. Schnable, "Chemically vapor deposited borophosphosilicate glasses for silicon device applications," *RCA Review,* Vol. 43, September 1982.
41. R. A. Carpio, "IR spectroscopy enhances dielectric film QC," *Semiconductor International,* August 1989.
42. J. E. Tong et al., "Process and film characterization of PECVD borophosphosilicate films for VLSI applications," *Solid State Technology,* January 1984.
43. W. Kern and R. Smeltzer, "Borophosphosilicate glasses for integrated circuits," *Solid State Technology,* Vol. 28, No. 6, 171, June 1985.
44. P. E. Riley et al., "Limitation of low-temperature low-pressure chemical vapor deposition of $SiO_2$ for the insulation of high-density multilevel metal very large scale integrated circuits," *The American Vacuum Society,* 1989.
45. C. G. Magnella and T. Ingwersen, "A comparison of planarization properties of TEOS and SIH4 PECVD Oxides," *VMIC,* 1988.
46. W. A. Pliskin, "Comparison of properties of dielectric films deposited by various methods," *The American Vacuum Society,* 1977.
47. A. C. Adams et al., "Characterization of plasma-deposited silicon dioxide," *J. Electrochemical Society: Solid-State Science and Technology,* Vol. 128, No. 7, 1981.
48. G. W. Hills et al., "Plasma assisted deposition and device technology: Interlevel dielectric consideration," *SPIE: Dry Processing for Submicrometer Lithography,* Vol. 1185, 1989.
49. H. Kotani et al., "Low temperature APCVD oxide using TEOS-ozone chemistry for multilevel interconnections," *IEEE IEDM,* 1989.
50. I. T. Emesh, "PECVD of silicon dioxide using tetraethylorthosilicate (TEOS)," *J. Electrochem. Soc.,* Vol. 136, No. 11, November 1989.
51. M. Simard Normandin et al., "Characterization of silicon oxide films deposited using tetraethylorthosilicate," *Canadian J. Phys.,* Vol. 67, 1989.
52. T. I. Kamins, "Structure and stability of low pressure chemically vapor deposited silicon films," *J. Electrochem. Soc. Solid State Science and Technology,* June 1978.
53. F. S. Becker and S. Rohl, "Low pressure deposition of doped $SiO_2$ by pyrolysis of tetraethylorthosilicate (TEOS)," *J. Electrochem. Soc. Solid State Science and Technology,* November 1987.
54. H. P. W. Hey and J. R. S. Machado, "An oxide step coverage comparison for different plasma CVD reactions (disilane/N2O, silane/N2O and TEOS/O2," *Proc. 175th Electrochem. Soc. Meeting,* Vol. 89, No. 1, May 1989.
55. M. J. Thoma et al., "A 1.0 μm CMOS two level metal technology incorporating plasma enhanced TEOS," *VMIC,* 1987.
56. W. S. Wu et al., "Characterization of $SiO_2$ films deposited by pyrolysis of tetraethylorthosilicate TEOS," *Journal de Physique,* Colloque C4, Supplement to Vol. 49, No. 9, September 1988.
57. S. Nguyen et al., "Reaction mechanisms of plasma and thermal assisted CVD of TEOS oxide films," *J. Electrochem. Soc.,* 137, 1990.
58. B. Ahlburn et al., "Advanced dielectric techniques for the fabrication of 16 megabit DRAM generation devices," *Proc. 3rd. Intl. Symp. on ULSI Sci. and Tech.,* ECS, Vol. 91-11, 1991.
59. S. Pennington, "An improved interlevel dielectric process for submicron double level metal products," *VMIC,* 1989.
60. A. N. Saxena and D. Pramanik, "Planarization techniques for multilevel metallization," *Solid State Technology,* October 1986.
61. A. Nagy and J. Helbert, "Planarized inorganic interlevel dielectric for multilevel metallization—Part 1," *Solid State Technology,* January 1991.

62. Y. Kup, "Planarization of multilevel metallization processes: A critical review," *SPIE Advanced Processing of Semiconductor Devices,* Vol. 797, 1987.
63. C. H. Ting, "Dielectric planarization processes for ULSI," J. Andrews and G. Celler (eds.), *Proc. 3rd. Intl. Symp. on ULSI Sci. and Tech., ECS,* Vol. 91-11, 1991.
64. F. Moghadam, "Planarization developments," *VLSI Multilevel Interconnection State of the Art Seminar,* 1992.
65. S. Sivaram et al., "Overview of planarization by mechanical polishing of interlevel dielectrics," J. Andrews and G. Celler (eds.), *Proc. 3rd. Intl. Symp. on ULSI Sci. and Tech., ECS,* Vol. 91-11, 1991.
66. P. Rentelin et al., "Characterization of mechanical planarization processes," *VMIC,* 1990
67. Technical note, "Planarization of thermal oxide sheet films," Westech Systems, Inc.
68. B. Davari et al., "A new planarization technique using a combination of RIE and chemical mechanical polish (CMP)," *IEDM* 89-61, 1989.
69. M. Thomas et al., "The mechanical planarization of interlevel dielectrics for multilevel interconnect applications," *VMIC,* 1990.
70. W. Patrick et al., "Application of chemical mechanical polishing to the fabrication of VLSI circuit interconnections," *J. Electrochem. Soc.,* Vol. 138, No. 6, June 1991.
71. Technical note, "Planarization of TEOS oxide interlayers over topography on 150mm wafers," Westech Systems, Inc.
72. V. Comello, "Planarization using RIE and chemical mechanical polish," *Semiconductor International,* March 1990.
73. J. Warnock, "A two-dimensional process model for chemimechanical polish planarization," *J. Electrochem. Soc.,* Vol. 138, No. 8, August 1991.
74. T. Daubenspeck, et al.," Planarization of ULSI topography over variable pattern densities," *J. Electrochem Soc.,* Vol. 138, 1991.
75. R. H. Wison, "Effect of planarization on VLSI metallization," *Semiconductor International,* April 1986.
76. H. Fritzsche et al., "An improved etchback process for multilevel metallization and its reliability results for CMOS devices," *VMIC,* June 1986.
77. P. E. Riley and E. D. Castel, "Planarization of dielectric layers for multilevel metallization," in L. B. Rothman (ed.), *Proc. Symp. Multilevel Metallization, Interconnection and Contact Technologies, ECS,* Vol. 87, No. 4, 1987.
78. S. K. Gupta, "Spin-on-glass films in multilevel IC interconnection," *Mat. Res. Soc. Symp. Proc.,* Vol. 108, 1988.
79. H. G. Tompkins and C. Tracy, "Tightly bound $H_2O$ in spin-on-glass," *J. Vac. Sci. Technol. B,* Vol. 8, No. 3, May/June 1990.
80. L. D. Molnar, "SOG planarization proves better than photoresist etch back," *Semiconductor International,* August 1989.
81. P. Pai et al., "Material characteristics of spin-on glasses for interlayer dielectric applications," *J. Electrochem. Soc.,* Vol. 134, No. 11, November 1987.
82. H. Watanabe and Y. Todokoro, "Submicron feature patterning using spin-on-glass image reversal," *J. Electrochem. Soc.,* Vol. 135, No. 11, November 1988.
83. C. Ting et al., "Spin-on-glass as planarizing dielectric layer for multilevel metallization," in L. B. Rothman (ed.): *Proc. Symp. Multilevel Metallization, Interconnection and Contact Technologies,* ECS, Vol. 87-4, 1987.
84. F. Whitwer et al., "Premetal planarization using spin-on-dielectric," *VMIC,* June 1989.
85. D. Yen and G. Rao, "Process integration with spin-on-glass sandwich as an intermetal dielectric layer for 1.2 μm CMOS DLM process," *VMIC,* 1988.
86. Y. Shacham-Diamand and Y. Nachumovsky, "Process reliability considerations of planarization with spin-on-glass," *J. Electrochem. Soc.,* Vol. 137, No. 1, January 1990.
87. B. Vollmer et al., "Characterization of field oxide transistor instabilities caused by SOG planarization," *Journal de Physique,* Colloque C4, Suppl. No. 9, Vol. 49, September 1988.
88. A. Schiltz, "Advantages of using spin-on-glass layer in interconnection dielectric planarization," *Microelectronic Engineering,* 5, 1986.

89. H. Kojima et al., "Planarization process using a multi-coating of spin-on-glass," *VMIC,* June 1988.
90. S. K. Gupta, "Spin-on-glass for dielectric planarization," *Microelectronic Manufacturing and Testing,* Vol. 12, No. 5, April 1989.
91. M. Kawai et al., "Interlayered dielectric planarization with TEOS-CVD and SOG," *VMIC,* June 1988.
92. M. Woo et al., "Characterization of spin-on-glass using fourier transform infrared spectroscopy," *J. Electrochem. Soc.,* Vol. 137, No. 1, January 1990.
93. S. Ito et al., "Application of surface reformed thick spin-on-glass to MOS device planarization," *J. Electrochem. Soc.,* Vol. 137, No. 4, April 1990.
94. N. Lifshitz et al., "Water related degradation of contacts in the multilevel MOS IC with spin-on-glass as interlevel dielectrics," *IEEE Electron Device Letters,* Vol. 10, No. 12, December 1989.
95. N. Parekh et al., "Plasma planarization utilizing a spin-on-glass sacrificial layer," in L. B. Rothman (ed.), *Proc. Symp. Multilevel Metallization, Interconnection and Contact Technologies,* ECS, Vol. 87-4, 1987.
96. B. Singh et al., "Deposition of planarized dielectric layers by biased sputter deposition," *J. Vac. Sci. Technol. B,* Vol. 5, No. 2, March/April 1987.
97. S. K. Ghandhi, *VLSI Fabrication Principles,* John Wiley & Sons, 1983.
98. L. M. Ephrath and G. S. Mathad, "Etching—application and trends of dry etching," in G. Rabbat (ed.), *Handbook of Advanced Semiconductor Technology and Computer Systems,* Van Nostrand Reinhold Co., 1988.
99. J. W. Coburn, "Plasma etching and reactive ion etching," *American Vacuum Society,* 1982.
100. J. W. Coburn and H. F. Winters, "Plasma etching—a discussion of mechanisms" *J. Vac. Sci. Technol.,* 16, 1979.
101. D. M. Brown et al., "Trends in advanced process technology—submicrometer CMOS device design and process requirements," *Proc. IEEE,* Vol. 74, No. 12, December 1986.
102. M. A. Jaso et al., "Etch selectivity of silicon dioxide over titanium silicide using $CF_4/H_2$ reactive ion etching," *J. Electrochem. Soc.,* Vol. 136, No. 12, December 1989.
103. P. Lee et al., "Chemical vapor deposition of tungsten (CVD W) as submicron interconnection and via stud," *J. Electrochem. Soc.,* Vol. 136, No. 7, July 1989.
104. B. Jucha and C. Davis, "Reactive ion etching of thick CVD tungsten films," *VMIC,* 1988.
105. R. S. Blewer, "Progress in LPCVD tungsten for advanced microelectronics applications," *Solid State Technology,* 1986.
106. S. Wu et al., "CVD tungsten and tungsten silicide for multilevel metallization," *IEEE Intl. Symp. VLSI Technol. Sys. Appl.,* 1989.
107. M. L. Yu et al., "Surface chemistry of the $WF_6$-based chemical vapor deposition of tungsten," *IBM J. Res. Develop.,* Vol. 34, No. 6, November 1990.
108. T. Ohba, "Multilevel metallization trends in Japan," *Advanced Metallization for ULSI Applications,* University of California, Berkeley, 1991.
109. H. Kotani, "Metallization for ULSI," *Extended Abs. 20th Conf. Solid State Dev. Matl.,* 1988.
110. M. E. Burba et al., "Selective dry etching of tungsten for VLSI metallization," *J. Electrochem. Soc.,* October 1986.
111. S. Sivaram and X. C. Mu, "Minimization of dimple size in tungsten filled contacts," *Extended Abstracts,* No. 248, ECS, October 1989.
112. T. Ohba et al., "Dimethylhydrazine surface cleaning and selective chemical vapor deposition of tungsten," *Advanced Metallization Applications,* University of California, Berkley, 1991.
113. R. W. Cheek et al., "Methanol passivation of $SiO_2$ for selective CVD tungsten," *Advanced Metallization Applications,* University of California, Berkeley, 1991.
114. W. L. Guthrie et al., "Chemical mechanical polishing to planarize blanket and selective CVD tungsten," *Advanced Metallization Applications,* University of California, Berkeley, October 1991.

115. E. L. Broadbent and W. T. Stacy, "Selective tungsten processing by low pressure CVD," *Solid State Technology,* December 1985.
116. N. I. Maluf et al., "Selective tungsten filling of sub-0.25 μm trenches for the fabrication of scaled contacts and x-ray masks," *J. Vac. Sci. Technol.,* B 8 (3), May/June 1990.
117. P. H. Singer, "Selective deposition nears production," *Semiconductor International,* March 1990.
118. G. Cameron and A. Chambers, "Successfully addressing post-etch corrosion," *Semiconductor International,* May 1989.
119. P. May and A. I. Spiers, "Dry etching of aluminum/TiW layers for multilevel metallizations in VLSI," *J. Electrochem. Soc.,* June 1988.
120. C. K. Hu et al., "Dry etching of TiN/Al(Cu)/Si for very large scale integrated local interconnections," *J. Vac. Sci. Technol.,* A 8 (3), May/June 1990.
121. J. W. Lutze et al., "Anisotropic reactive etching of aluminum using $Cl_2$, $BCl_3$ and $CH_4$ gases," *J. Electrochem. Soc.,* Vol. 137, No. 1, January 1990.
122. S. Gupta et al., "Materials for contacts,barriers and interconnects," *Semiconductor International,* October 1990.
123. J. K. Elliot, "Current trends in VLSI materials," *Semiconductor International,* March 1988.
124. M. A. Nicolet, "Diffusion barriers in thin films," *Thin Solid Films,* 52, 1978.
125. T. Nishida et al., "Multilevel interconnection for half micron ULSIs," *VMIC,* 1989.
126. Y. Arita et al., "Deep submicron Cu planar interconnection technology using Cu selective chemical vapor deposition," *IEDM,* 1989.
127. H. Yamamoto et al., "Reliable tungsten encapsulated AlSi interconnects for submicron multilevel interconnection," *IEDM,* 1987.
128. J. E. J. Schmitz, *Chemical Vapor Deposition of Tungsten and Tungsten Silicides,* Noyes Publication, 1992.

Chapter

# 3

# Manufacturing

## 3.1 Introduction

This chapter deals with manufacturing issues associated with integrated devices that use multiple levels of metallization. These issues are quite different for integrated circuits (ICs) that need only a single level of metallization. The fabrication issues are complex in nature and this complexity lies not only in the multilevel technology but also in the manufacturing methodology. Even though a company may have a good product design and a process, the real litmus test in seeking a competitive market advantage and share lies in a company's ability to manufacture the products in high volume at the lowest possible cost. This chapter is divided into four segments: manufacturing challenges, process integration, yield analysis, and defect reduction. These topics are examined here in the context of manufacturing advanced ICs, specifically those that use multilevel interconnects.

Manufacturing plays a critical and central role in IC fabrication. The challenges are multifaceted, ranging from advances in design and process to manufacturing practices. The manufacturing challenges may appear simplistic and mundane, but the goal is to take an advanced technology and produce products in large quantities that are safer, faster, cheaper, better, and easier to make than those of the competition.

Each process must be developed taking into account manufacturing issues. One of the process integration challenges is in creating a process flow that is easily transferable to manufacturing. The requirements and goals of process integration vary from the process development to production.

Yield analysis is one of the key indicators that conveys the degree of success of the process as a whole and the manufacturing technique in particular. The yield loss of a product due to wafer processing is examined, and yield models that help calculate and compare different product yields are also presented in this chapter.

Finally, continuous improvement in both the process and manufacturing procedures drives the defects down, resulting in an increase in profitability. The focus here is on defect reduction methodologies. With such a mix of challenges, the fabrication of ICs is an exciting field of study.

## 3.2 Manufacturing Challenges

Manufacturing is an exciting field where people, equipment, and technology have to be managed for fabrication of products. The challenges this offers are many.[1–3] The focus of this chapter is on the manufacturing challenges that multilevel interconnect technology presents. The sequence of steps that are needed to fabricate multilevel interconnects poses the potential for manufacturing bottlenecks and yield depression due to increased defect levels. One of the reasons is because of the multitude of repetitive steps that are required to fabricate interconnects. Effective process transfer methodologies from development to production, coupled with sound manufacturing practices, can overcome most of the problems associated with the process implementation. The *backend* (BE) of a process flow may be defined as the sequence of steps starting after the transistors are formed. That is, the BE of a process flow comprises multilevel interconnects—deposition, planarization, etching, etc. The *frontend* (FE) may be defined as all steps starting from silicon start to the formation of transistors—active and isolation areas, gate definition, etc.

Overcoming yield problems associated with multilevel interconnects is an extremely important activity because the value of silicon wafers increases as they progress toward the BE. Once wafers leave the fabrication site and go to the assembly site, they increase in value as they are placed in a package and tested. Of particular interest are defects and process deviations caused by interconnect and dielectric steps. These defects and deviations may lead to the loss of a significant number of wafers which have accumulated a potentially high market value. The graph shown in Figure 3.1 shows the cost of identifying defects at various points in the process flow. Obviously, identification and scrapping of defective material at the start of the process is much less expensive than at the end, which may involve analyzing customer returns for defects and recalling the defective products. It pays (lowers cost) to identify and eliminate defects at the source or close to it. As technologies push the level of device integration even higher with use of multiple levels of interconnects, the cost per wafer starts rising. To reduce the cost per die, the migration to larger wafer

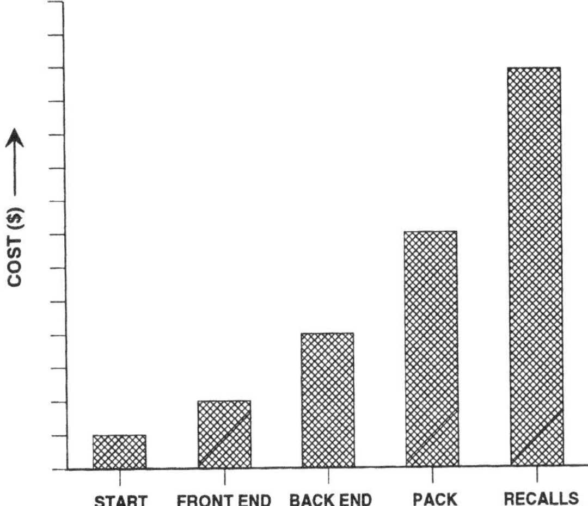


**Figure 3.1** Relative cost of indentifying defects at some locations in a wafer fabrication flow.

sizes is becoming necessary as more die can be printed on a wafer. However, increased wafer size comes with its own issues.

### 3.2.1 Manufacturing methods

There are two methods, proactive and reactive, for achieving manufacturing success. These are not alternatives that one can choose from, but necessities for good manufacturing. Figure 3.2 shows the flow of information from both proactive and reactive activities into the manufacturing process. Each of these methods is a part of the feedback system that is constantly aimed at improving designs and processes so that high-volume manufacturing is feasible with maximum yields.

In the proactive method the manufacturing concerns are incorporated at the design and process development stage. The manufacturing issues are resolved though a constant and regular feedback loop between design, process, and manufacturing organizations in an IC fabrication plant. This interactive learning process enables designs and processes to be developed that can be quickly put into high-vol-

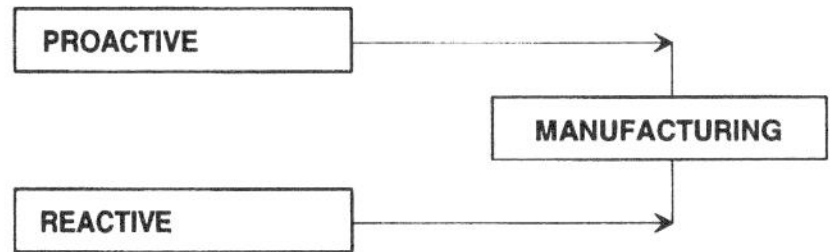


**Figure 3.2** Proactive and reactive modes of information flow in a manufacturing system.

ume production. It is only through high-volume production that two important benefits may be realized. The first benefit is an increase in the cycles of learning that may not have been possible during process development due to long throughput times. These cycles of learning shorten the time to understand and fix potential yield limiters. The second benefit of increased volume of production with high line and die yields is the opportunity to penetrate the market ahead of the competition and recover most of the developmental costs in the early phases of product introduction. A successful company will do this before the product transitions from a proprietary to a commodity product.

The reactive mode of operation involves addressing systematic and random excursions that degrade the yield, which is a function of manufacturing operations. Depending on the magnitude and nature of the problem, the manufacturing process may be halted for some time pending a resolution of the issue. Reaction to excursions can made easier by using inline and end-of-line monitor data that can flag early indications of the onset of a problem. Experience gained from successfully addressing manufacturing issues may be captured in cause and effect diagrams, troubleshooting, checklists, etc.

### 3.2.2 Customer concept

Manufacturing of submicron ICs is well placed in a pivotal role; on one hand we have the customer and on the other we have the company's own profitability and survival at stake. As manufacturing technologies increase in complexity for submicron geometries, the customer needs to be placed in a central position and educated about key operations and products. Such an education helps foster a better relationship with the customer and makes them a partner in the joint progress of both the customer and vendor companies. Figure 3.3 shows a possible scenario of the relationship of various groups in an IC manufacturing

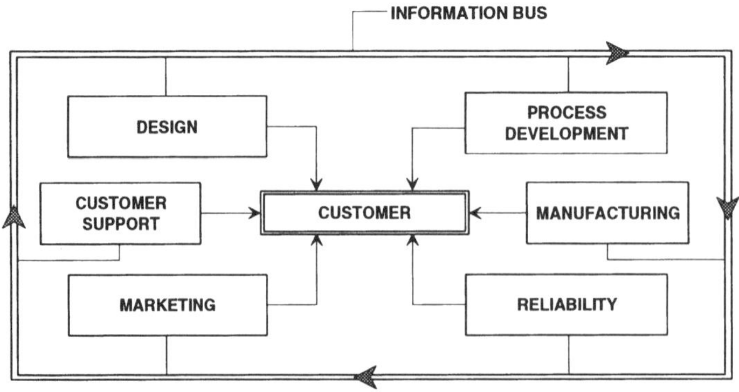


**Figure 3.3** A possible configuration of customer interface system with key areas within the vendor organization.

company to the customer. This is not to suggest that the customer should deal with a host of organizations. This figure is meant to illustrate that each organization should comprehend the needs of the customer, be it internal or external. Each group should constantly understand the issues of the other groups via an information path that is available for data and information collection and dissemination. Such a relationship is much needed in an application-specific integrated circuits (ASIC) fabrication environment where rapport with the customer is vital to the successful fabrication of the products that meet the customer's needs. In commodity product [such as dynamic random access memories (DRAMs)] fabrication, the relationship may not be as close, but it is still important to ensure that the products and services are performing to standards set by the customer.

Manufacturing directly impacts the profitability of both the customer (on-time delivery of products at competitive cost) and the company fabricating the ICs itself. In Figure 3.3, manufacturing, along with other groups, has a key role to play in enhancing the customer relationship. There are many ways to provide good support to customers; one of them is through education of the fabrication process as a whole. Education brings a better appreciation of the issues facing the vendor.

### 3.2.3 Technology trends[4]

On a macroeconomic scale, the market is continuously emphasizing the decline in IC unit cost. This is especially true in the case of DRAM fabrication where high-volume manufacturing is a significant leverage item in reducing unit cost. Unit cost is further driven down by using larger-diameter wafers and increasing line and die yields. Simultaneously, new generation design and process technologies are being developed to increase the packing or integration density (number of transistors per chip) so that faster and smaller chips may be manufactured in larger quantities.

**Packing density.** The packing density of transistors has been constantly increasing, primarily due to the need to enhance the application scope of the IC. Figure 3.4 shows an increasing trend of transistor packing density. Defining the interconnects for all these transistors is a challenging task. The interconnect layout must be done using multiple levels of metallization as the die size using a single layer of metallization would be large and more adversely affected by high defect densities. The critical question is, how can we ensure that our parametric and reliability testing are adequate for predicting the functionality and reliability of the millions of transistors in the die? The answer lies more in the methodology of process control and continuous improvement and less on additional testing. The increase in the pack-

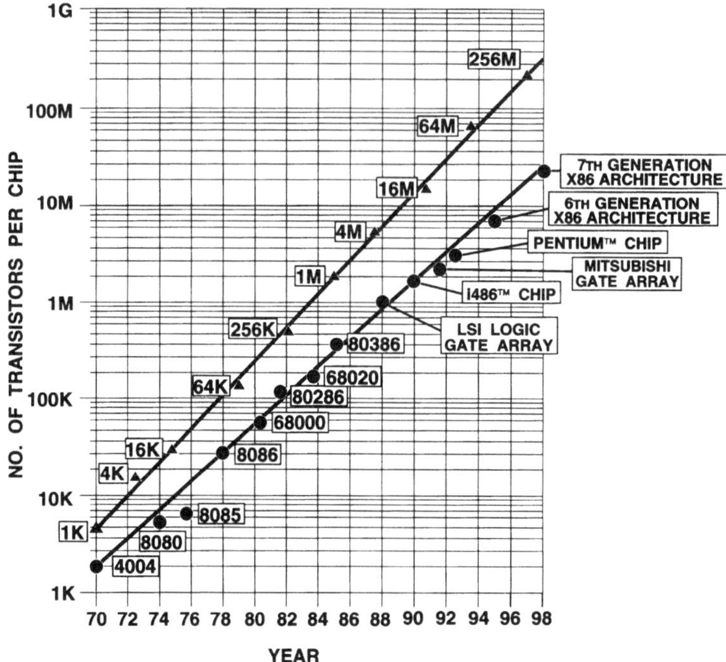


**Figure 3.4** Transistor packing density trend.[4] (*Courtesy of Integrated Circuit Engineering Corporation.*)

ing density requires new process technologies with ever-decreasing feature sizes in order to extract the full benefit of the product design.

**Feature size.** The downward scaling of device dimensions enhances the speed and power performance of devices and also allows the increase in the transistor packing density. Figure 3.5 shows the trend of the feature size over time. This diagram illustrates the current production and experimental resolution in terms of device fabrication and prototyping, respectively. Feature size reduction pushes the process technology from one generation to the next. This means that significant process enhancements are required to transfer the designs to silicon (Si). Feature size reduction is limited by fine pattern formation of the lithography tools and the ability to develop multilevel interconnect systems. The process technology for each generation is dictated by the speed of development on two fronts—transistors and interconnects. As shown in Figure 3.5, the formation of devices in the sub-quarter-micron regime has been demonstrated. Besides the optimization work for transistor performance, one of the key requirements for high-volume manufacturing is the development of interconnect technology. The reduction in feature size presents a whole new challenge toward the definition of multilevel interconnects. Once the

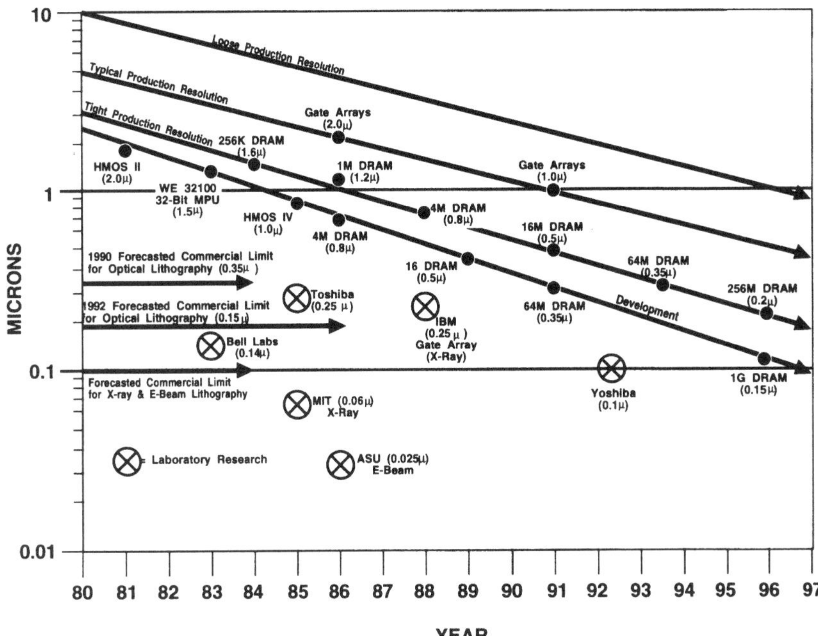


**Figure 3.5** Device feature size trend due to scaling.[4] *(Courtesy of Integrated Circuit Engineering Corporation.)*

process has been developed for both transistors and interconnects, the manufacturing challenge is in producing wafers with high yields, as smaller defects may become potential yield killers.

**Die size.** In general, the die size has been increasing with time and keeping pace with the increase in the transistor packing density. This trend is illustrated in Figure 3.6, which shows the die size increase over time for both the microprocessor and the memory devices. These two device technologies have kept pace with each other. Die size increase is inevitable as transistor packing density increases with each new device generation. For a particular device generation, the die size may be reduced to increase the number of dice per wafer. This is done by design shrink and by using the next generation of process technology, which enables the fabrication with smaller feature sizes. The benefits of increasing die size are packing density increase, enhanced circuit layout, and better circuit function.

However, from a manufacturing perspective, die size increase presents several problems as highlighted in Figure 3.7. These problems may be overcome by developing a more advanced technology that may compensate for these side effects. The main problem of die size increase is elevated levels of defects (higher probability of defects falling on die) and hence lower yields.

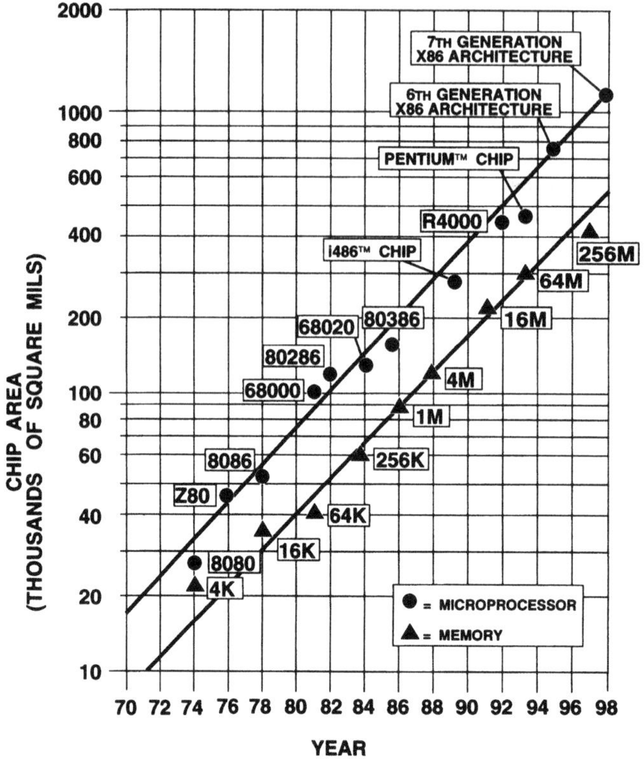


**Figure 3.6** Die size trend.[4] (*Courtesy of Integrated Circuit Engineering Corporation.*)

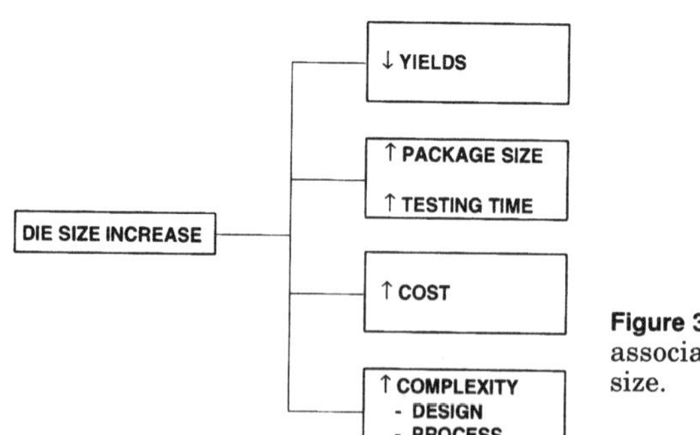


**Figure 3.7** Benefits and concerns associated with increase in die size.

**Wafer size.** To reduce manufacturing costs, an increase in wafer size is needed as more dice may be fabricated on a single wafer. The current trend for most advanced IC manufacturers is to transition from 150- to 200-mm wafers. Some of the processing challenges resulting from increasing wafer diameter are modulated by equipment issues in terms of ensuring tight uniformity control across the wafer. Wafer handling and inspection are also more difficult on large-diameter wafers.

### 3.2.4 Process complexity

The manufacturing operation may be divided into two groups, one for transistor formation and the other for interconnect formation. It is interesting to compare these two groups and attempt to understand the complexities of each. The issues listed are the output parameters and concerns that the manufacturing operation must ensure in order to meet process and product specifications. Figure 3.8*a* and *b* compares the salient steps in both transistor and interconnect formation processes.

The transistor formation steps are generally batch processing involving large load sizes that may cause major line yield loss should there be a processing problem. On the other hand, some of the interconnect formation steps are single-wafer operations. The real differences between the two groups are in planarization, metallization, and etching. The comparison may be further broken down into materials, temperature, and cleans.

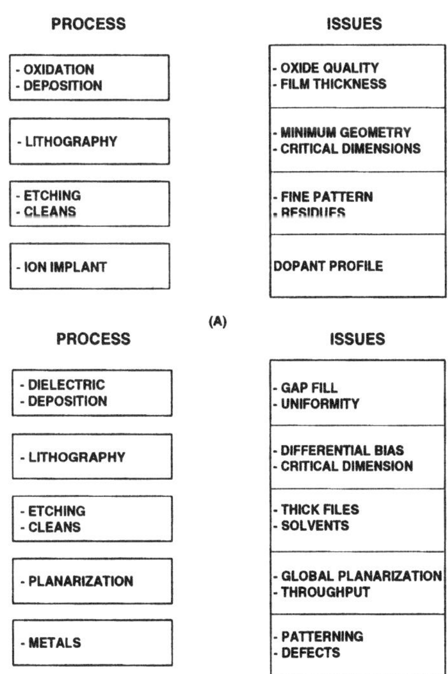


**Figure 3.8** Comparison of key steps in (*a*) transistor formation and (*b*) interconnect formation.

**Materials.** In the transistor formation process blocks, the process complexity lies in precise control of dopants in the silicon to form the requisite wells and junctions by optimizing the oxidation, diffusion, and implant steps. The thickest film [in a metal-oxide semiconductor (MOS)] using thin film deposition techniques is the polysilicon layer, and with oxidation it is the field oxide. Even though the feature sizes are small, the aspect ratios are not as large as in the BE or interconnect formation steps so the common problems of filling high-aspect ratio gaps with thin films do not arise. The primary materials that are used in MOS transistor formation are Si, silicon dioxide ($SiO_2$), and silicon nitride ($Si_3N_4$). Compare this with the number of materials that are used for the formation of interconnects—several types of dielectrics [oxides, nitrides, spin-on-glass (SOG), and polyimides] and metals [aluminum (Al), tungsten (W), titanium (Ti), etc.]. Each of these materials has different electrical and physical properties that make process development and control a challenging task. The management of the materials is therefore critical for the manufacture of ICs with multiple levels of interconnects.

**Temperature.** High temperature is a major constraint in all of the interconnect formation steps as Al and its alloys are widely used as interconnects. Pure Al melts at 660°C and this temperature starts decreasing as Al alloys with Si and/or copper (Cu) are formed. The Al/Si eutectic temperature is around 577°C and the solubility of Si in Al rises as the temperature increases.[5] Therefore, Al-Si alloys are used to prevent junction spiking and the post aluminum temperatures are restricted to around 400°C. An increase in temperature also causes the Al to relieve stress in the form of hillocks. Increase in temperature induces a stress in the metal due to a difference in expansion coefficient of the metal and the silicon. Hillock density may also increase with an increase in impurity content (nitrogen). Temperatures in the dielectric deposition are also controlled with the usual operating temperature around 300°C to prevent the formation of hillocks. Unlike the interconnect steps, transistor formation steps (oxidation, diffusion, and deposition) are done at elevated temperatures. Temperature effects are well characterized and understood for thin films and dopant distribution in silicon.

**Cleans.** In addition to the number and nature of materials, the wet chemical treatments are different for each group. For the transistor formation group, acidic cleans do a reasonable job in removing resist residues and other contaminants. However, for the interconnect cleans, nonacidic cleans are necessary in order to prevent metal attack. Organic resist strippers are widely used in addition to plasma ashers to strip resist and clean the surface. These cleans are not as efficient as acidic ones that are used for the transistor formation steps.

These differences present a different set of process integration challenges for the two groups. However, from a manufacturing perspective, the issues between the two groups are similar except when it comes to metallization, which is a critical operation in the manufacturing flow. The manufacturing issues stem from the repetitive nature of the interconnect processes. The financial value of the wafer increases with each interconnect level. Misprocessing at any of these stages causes a significant loss of time and money.

### 3.2.5 The metal loop

As the interconnect levels increase, the metallization complexity in terms of manufacturing logistics increases too. Figure 3.9 breaks down the actual metallization steps that are required to form one level of interconnects in a typical multilevel interconnect system. A process that has three levels or four levels of interconnects consists of several metallization steps. In Figure 3.9 there are six steps that form one interconnect level. So for a three-level multilevel interconnect system, there are 18 metallization steps. Inclusion of source/drain and polysilicon metallization will increase the total number of metallization steps still further. This assumes that the metallization scheme is Al-based, as barrier and shunt layers are needed to make interconnects robust. Multiple metallization steps could constrain the flow of material. The bottlenecks may be equipment or process related. Some relief may be obtained by using common metallization steps for various applications. For example, the adhesion layer in a W plug process may be TiN and these films may also be used as barrier and antireflective layers. This enables the same equipment set and process to be used over and over again. Such a methodology in process development makes manufacturing more efficient.

Line balance is essential to ensure a smooth flow of material through the manufacturing line. In a short-cycle time environment,

| LAYER # 6 : ANTI-REFLECTIVE LAYER |
|---|
| LAYER # 5 : SHUNT LAYER |
| LAYER # 4 : PRIMARY CONDUCTOR |
| LAYER # 3 : BARRIER METAL |
| LAYER # 2 : CONTACT/VIA METAL |
| LAYER # 1 : ADHESION LAYER |

**Figure 3.9** Identification of components in a typical metal interconnect.

line balance requires that the various operations be synchronized, as they normally run at different speeds. The interconnect steps are a mixed bag of single-wafer and batch process steps which are fed by the FE operations which are mostly batch. The situation is further compounded by a multitude of repetitive non-value-added steps such as inspections for defects, and machine and process monitors in the metallization steps. Therefore, along with the reduction in cycle time, a careful evaluation of the non-value-added steps is required. An effective utilization of the monitors and inspections is needed to not only obtain all the necessary information for checking the status of the process but also to improve the flow of material through the line.

### 3.2.6 Cycles of learning

In both development and high-volume manufacturing the benefit of shorter cycle time can not be overstressed. It is one of the most important manufacturing indicators which establishes success of the manufacturing operation. Short cycle times increase the number of cycles of learning or information turns and therefore provide an opportunity to react to process excursions quickly. What is cycle time?

*Cycle time* or *throughput time* is the time required for material to move from the start to the end of the flow. It is a measure of how quickly the work-in-process (WIP) moves though the line. Each piece of equipment has its own throughput time and the sum of these times plus the material handling times plus inspections make up the theoretical throughput time of the manufacturing line. In practice, however, cycle time is influenced by line yield at each step, the WIP levels, and manufacturing practices. Cycle time may be expressed in by a simple formula, as shown in Equation (3.1).

$$\text{Cycle time} = \frac{\text{WIP inventory}}{\text{processing speed}} \tag{3.1}$$

The advantages of having short cycle times are many and some of the salient ones are given in Figure 3.10 (a complete listing of the advantages is beyond the scope of this book).

**Customer.** This consists of benefits 1 and 6. A decrease in the cycle time will help the customer by predicting the product delivery schedule more accurately and by providing a quick response. With shortened cycle time, problems become visible and reaction to fix these can be started in a timely manner.

**Profits.** Benefits 2, 4, and 5 may be considered to be part of this group. One immediate advantage of achieving short cycle time is the reduction in inventory and its carrying costs. Just-in-time manufacturing concepts may then be applied to further leverage the manufac-

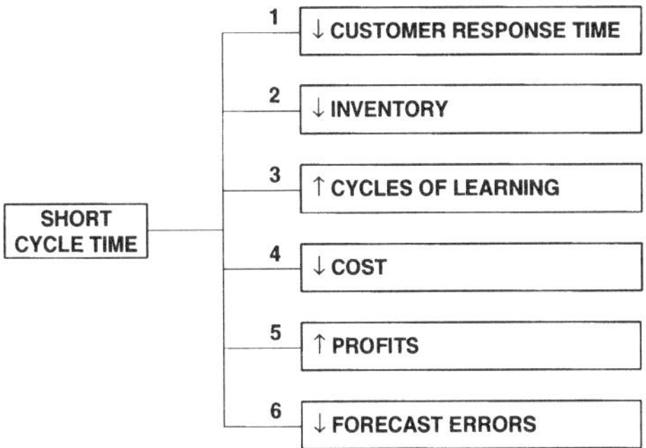


**Figure 3.10** Benefits of short manufacturing cycle time.

turing operation in securing the gains of reduced inventory levels. The relationship between cost and cycle time needs to be explored a little further. As we continue to decrease cycle time we are able to reduce costs to a point. There is a point when further reduction in cycle time begins to increase cost and is no longer attractive.

**Learning.** One of the key benefits of short cycle time is the opportunity to increase the cycles of learning. This benefit is vigorously sought by engineering to understand the cause and effect of various processes and to determine the effectiveness of process improvements. The concept of a *cycle of learning* (CL) is very important to yield improvement. A cycle of learning may be defined as the elapsed time from a process change until the results of the change are analyzed. Expressed mathematically, cycles of learning (CL) may be defined as

$$\mathrm{CL} = \frac{\mathrm{WD}}{\mathrm{CT}} \tag{3.2}$$

where WD = workdays in a year
CT = cycle time in days

Increasing the cycles of learning in a manufacturing operation provides the opportunity to identify problems and rectify them faster. With the increasing sophistication of processes, the number of steps is increasing and this tends to make the cycle times longer.

Cycles of learning may be increased in two ways, first by reducing the overall cycle time and second by sampling the process at regular intervals and reacting to the data before a problem reaches end of line. Let us take a closer look at CLs as a result of increased sampling by looking at a multilevel metal process technology flow with n layers of metal. Figure 3.11 shows a feedback loop in a n-level metal process

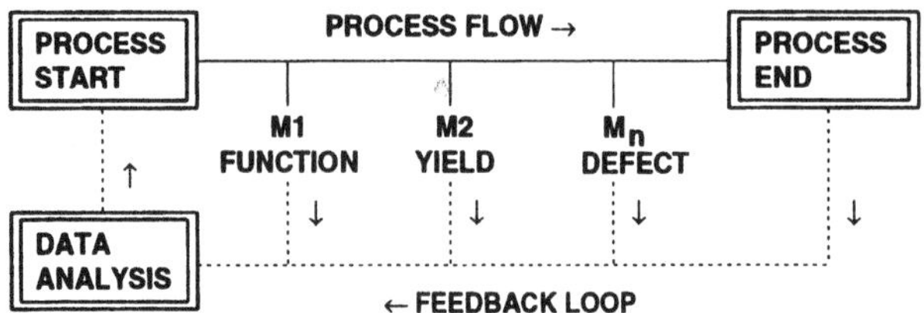


**Figure 3.11** A process health sampling feedback system.

flow. In this figure, the process begins with the Si wafers being started for the various products. After completing their inline processing, the Si wafers move to test where die yield is determined. As this process uses multilevel interconnects, it may be segmented into several subprocesses and the learning activity may be tailored for each subprocess. In a multilevel metal system, let the final metal layer be $M_n$. The process flow may be sampled at $M_1$, $M_2$, and $M_n$ interconnect levels before the lot reaches test for yield evaluation.

Transistor functionality may be tested once the first-level interconnects ($M_1$) are made. Testing samples at this location provides a quick and early indication of the transistor functionality. This is a good location to check for contact resistances, threshold voltages, etc. To speed up the learning process, special short process flows may be created that terminate at $M_1$. The short loop process flow can take advantage of inherent low cycle time of the manufacturing line, and further increase the cycles of learning. Experiments may be defined on these short loops in order to characterize process specification windows associated with some of the critical processing steps.

The above concept may be extended to other levels of interconnects. To check circuit functionality in some devices, the wafers may be processed on a different flow and terminated after the second interconnect level ($M_2$) has been defined. This provides an early indication of any yield problems. This is an excellent way to check the yield impact of a process change; for example, on static random access memory (SRAM) yield vehicles that may be fabricated with two layers or more of metallization. Low-yield analysis may be performed on such material and valuable information on the failure modes may be obtained well before a fully processed lot with all levels of interconnects makes it to the end-of-the-line. In addition to this, separate process flows may be created to check the continuity, via resistances, etc. As each interconnect level is similar to the previous one in a multilevel interconnect system, the defect generation sources may be studied if the wafers are inspected at the last interconnect level and then sent on for end of line testing. Inspection for the defects after the final layer does provide a cumulative defect count. This provides an opportunity to correlate the visual defect density to the electrical defect density (yield), which enables fast feedback for reaction. All wafers are tested twice: First, to check the electrical parameters and second, for the circuit functionality. Information from these two tests helps determine an appropriate failure analysis, if needed. The bottom line is to increase the cycles of learning through aggressive management of cycle time reduction and creative process sampling and monitoring.

## 3.3 Process Integration

Process integration is one of the key functions in the process development and manufacturing of ICs. There are many integration aspects when fabricating ICs with multilevel interconnects. Process integration focuses on the health of the process as it migrates from development to high-volume manufacturing. This process is highlighted in Figure 3.12 where areas of focus are listed in blocks.

### 3.3.1 Development

During the process development phase, the process integration is concentrated on defining process flows, characterizing new processes, studying process interactions and addressing systematic process problems. An example of process interaction is the stress in thin films which must be carefully managed. Mismatch of stresses between films may lead to film delamination or other catastrophic failure. Once interactions are understood, the process must be controlled within specification limits that may be based on product requirements and on the process variability.

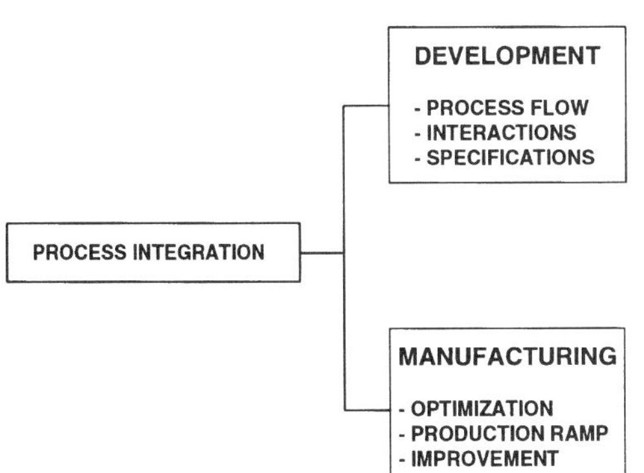


**Figure 3.12** Process integration challenges during process development and high-volume manufacturing.

### 3.3.2 Manufacturing

As shown in Figure 3.12, key integration issues in high-volume manufacturing are different from process development. Successful transfer of a process from development to high-volume manufacturing depends on its maturity and the discipline in technology transfer. Process optimization and improvement may be done after process transfer and are required for the reasons given below.

**Machine matching.** Several pieces of equipment may be used for any process step. It is absolutely essential to ensure that each of these machines are matched so as to obtain a tight distribution of the output parameters. The objective is to drive output electrical parameters closer to target for all machines and reduce the variation between them.

**Process envelope.** Process envelope or window is the region in which the process may operate without jeopardizing products. During process development, process window experiments can be used to characterize a process. However, during high-volume fabrication the process envelope may change due to volume effects. As an example of volume effect consider a process flow that may have been developed with a short time delay between borophosphosilicate glass (BPSG) deposition and densification and reflow, avoiding crystal formation. In volume production, the characterized time between BPSG and densification may not be adequate. Cycle time and manufacturing practices may require a wider time limit between processes. Therefore, reevaluation of the process envelope to provide a larger time window for operation may be needed if the process must be improved and made suitable for high-volume manufacturing. In some cases, process improvement is possible and in others it is difficult. It is important not to confuse process optimization (centering the process and verifying the process envelope) with re-engineering/development of a process. If required, reengineering/development of a process may enhance the yield or reliability of the product, and should be done carefully. In high-volume manufacturing, the risk to the product is far greater than in development mainly due to the larger amount of material that is being processed. Process problems due to inadequate characterization in development will cause a significant loss in material, time, and revenue in production.

**Problem investigation.** The process integration function in a manufacturing environment is more investigative in nature as opposed to discovering/inventing in process development. Investigative engineering requires a thorough understanding of the process and process interactions. The key is in problem solving. Failure to quickly and successfully address production issues may result in downtime and substandard product. The lessons learned from various problem solving opportunities enable designers and next-generation technology developers to improve new product designs and processes. Through investigative engineering both proactive and reactive components of manufacturing may be exercised. The reactive part translates into a process fix while the proactive part may translate into a modification of design or process. It is for this sole purpose that the dialogue between designers, product, process, and reliability engineers is extremely important.

What tools are needed for investigative engineering? There are many—electrical test structures, computer tools, inline process monitors, reliability tests, and customer feedback. Process monitors are widely used to monitor the equipment and the process. For example, these monitors may include thickness or critical dimension checks. Process monitors may or may not flag a device problem, but electrical tests do provide some level of confidence that product has met specifications as it leaves the factory.

There may be a tendency to increase the number of monitors that check various conditions of the process and equipment. But this poses a problem as monitor wafers are processed at the expense of product wafers. Moreover, the results of the monitors may gate movement of product wafers and hence increase cycle time, which results in reduced CLs. It is important to evaluate the monitors across the process and optimize their function as the factory ramps to process a larger volume of product.

## 3.4 Yield Analysis

The aim is to make high-yielding wafers and reliable dice. In an ideal situation the yield of a properly fabricated wafer would be 100 percent. In practice, however, this is not the case and the yield varies depending on the maturity of the process from a few percentage points to near 100 percent. The mechanisms of yield loss are many and the focus of this section is to examine how multilevel interconnect processing effects yield.[6–23]

### 3.4.1 Yield components

The causes of yield loss fall into two main defect categories: systematic and random. During process infancy, systematic defects are the main contributors to yield degradation. As the process matures, the random component becomes the focus of defect reduction as systematic defects decrease. Figure 3.13 shows defect reduction over time for any process. This may be attributed to progress in addressing the systematic and random defects that degrade yield. Also shown

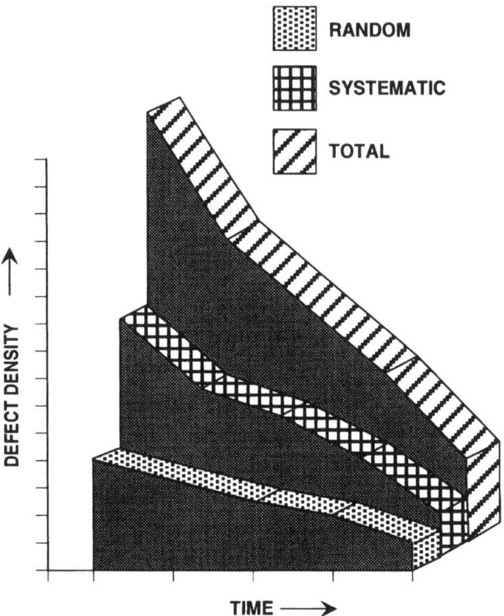


**Figure 3.13** Total, systematic, and random defect density trend.

in this figure are two other curves, one representing the systematic component and the other the random. To a large extent, systematic yield degradation is addressed during the process development stage. The process should be transferred with a minimum of systematic defects. However, as process transfer from development to manufacturing may be expedited by market pressures, the burden of reducing both types of defect modes lies with the manufacturing organization, when production is being ramped to meet demand. Once the majority of systematic defects are addressed, yield and reliability are limited by the inherent random defect component of the process. The random component is always present even in mature and high-yielding processes. At some time after the process has demonstrated maturity (high line and die yields with no reliability issues) a decision has to be made whether to invest resources to bring defect levels further down or to switch to the next-generation process which may offer a yield advantage. With each new generation of process technology, defect specifications for selecting a piece of equipment are tightening.

**Systematic yield loss.** Figure 3.14 shows a wafer with a grid of dice that have been tested. On this grid, "r" represents dice that have failed due to the presence of a random defect and the square dot pat-

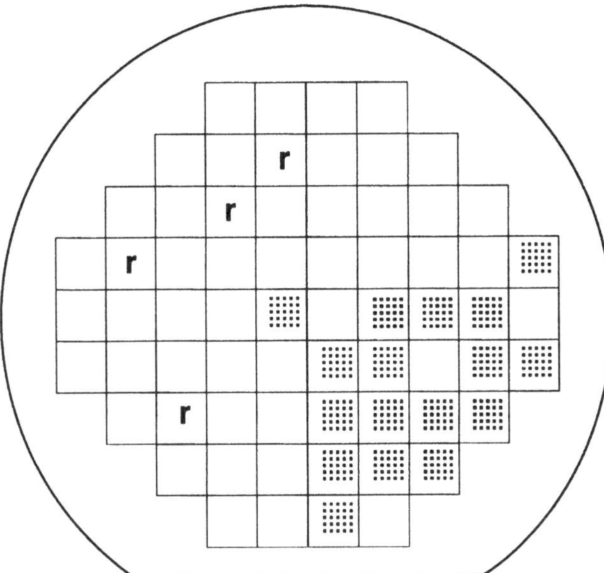


Figure 3.14 An example of a wafer map after completion of yield testing showing die failure due to random and systematic problems.

tern represents dice that have also failed, but due to a systematic problem. Large failing areas suggest systematic problems that may be due to some process step or an interaction of steps.

An example of systematic yield loss is the mismatch of uniformities of adjacent process steps. Several systematic yield problems may be attributed to noncomplementary thin film deposition uniformity profiles and the subsequent etch uniformity profiles, even though for the individual process steps the uniformity may be in the specification. Consider a thin-film deposition process that exhibits thickness uniformity discrepancy between the center and the edge of the wafer, with the center being thinner. Consider also the subsequent plasma etch profile that has a high etch rate in the center as compared to the wafer edge. When the two processes are combined in a process flow with the etch process following thin film deposition, the remaining film thickness will be thin in the center as compared to the wafer edge due to overetching. This could result in severely thin areas in the center of the wafer that may lead to processing problems, as shown in Figure 3.15, where affected die are denoted by the square dot pattern. Several combinations of uniformity profile exist and their interactions may or may not produce an adverse yield condition. Such a systematic problem may be fixed by adjusting the uniformity profiles of upstream and downstream process steps so that they do not

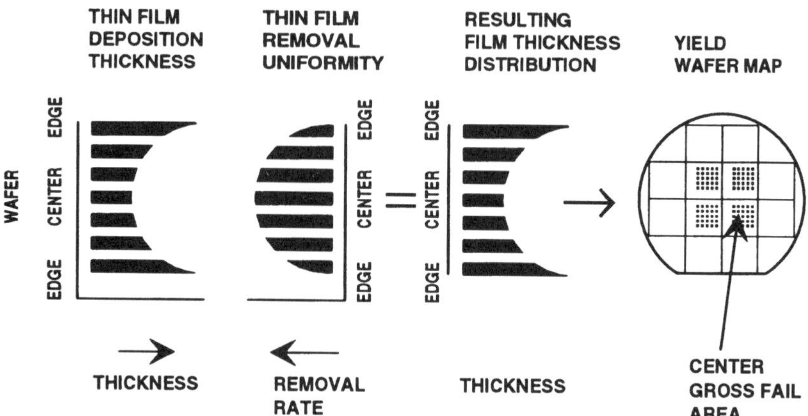


**Figure 3.15** Impact of destructive interference of deposition and etch process uniformities on yield.

interact in a negative way to create an overetched or underetched situation, as described in the above example.

**Random yield loss.** Yield loss is due to a random occurrence of defects, then failure will be randomly distributed across the wafer as shown in Figure 3.14. Random yield losses are more difficult to address and isolate to a particular processing step. Several models exist to evaluate yield loss due to the random yield component. The impact of multiple types of random defects on the yield are also part of some of the models.

### 3.4.2 Yield loss

The cause and effect diagram for yield loss is shown in Figure 3.16. This diagram also highlights the impact of multilevel interconnect processing on yield. To gain a better understanding of the BE issues that could cause yield degradation, the cause and effect diagram has been divided into four groups. As will be more evident in the defect sections, classification of defects into systematic or random is not always clear-cut. But, irrespective of this difficulty, the cause and effect diagram highlights the various defects that degrade yield in the BE of the process. Particles are a real problem and effect all aspects of processing from the beginning to the end of the process flow. It is to be hoped that when a product is ready for high-volume manufacturing, most of the product issues will have been resolved. In some cases, problems are discovered after a product has been put into production. Poor manufacturing practices may cause yield losses.

Yield degradation may be classified as either catastrophic or parametric in nature. *Catastrophic failures* are those that make the cir-

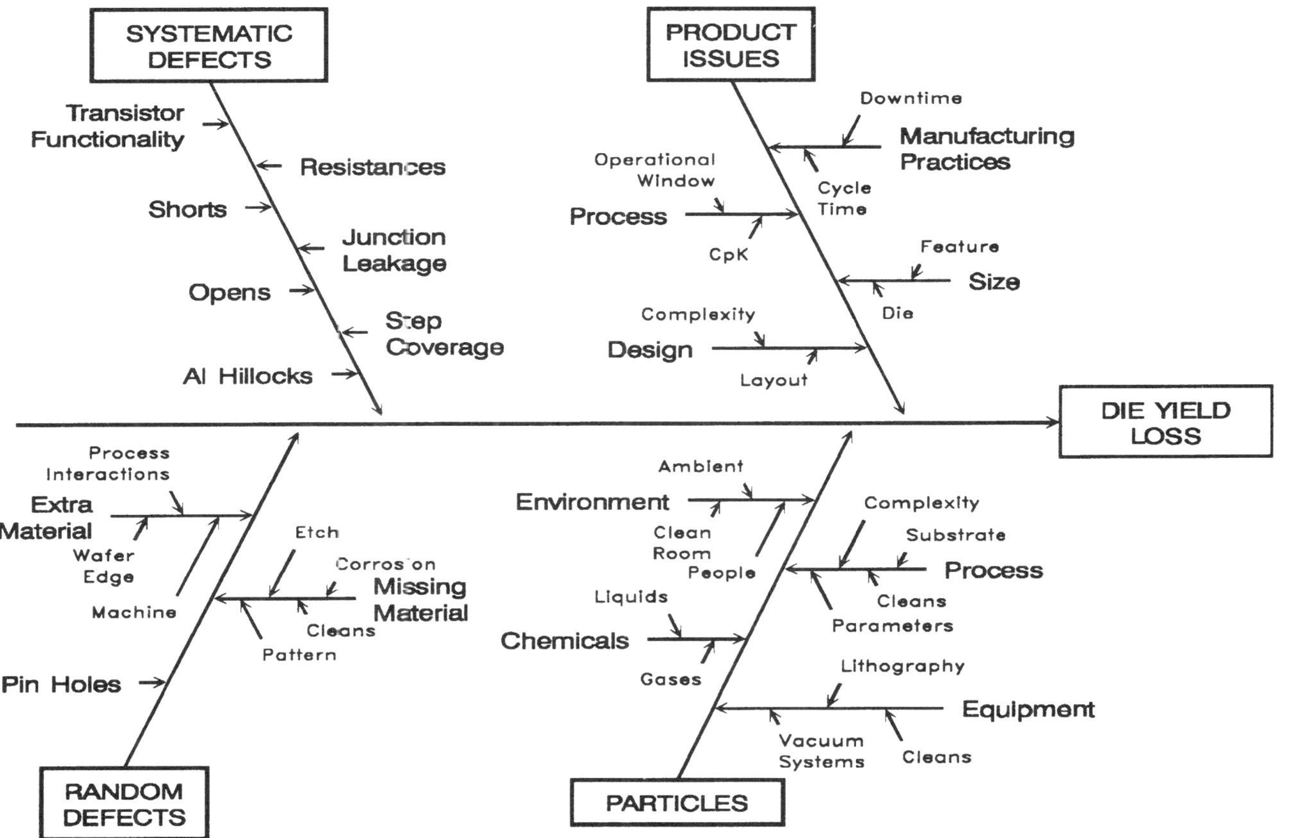


Figure 3.16 Cause and effect diagram for yield loss.

**TABLE 3.1 Catastrophic and Parametric Failures**

| Defects | Catastrophic failure | Parametric failure |
|---|---|---|
| Systematic defects | Opens<br>Shorts | Step coverage<br>High resistances |
| Random defects | Extra material<br>Missing material | Particles<br>Pinholes |

cuit fail completely and the yield is zero or close to it. *Parametric failures* are those that do not drop the yield significantly, but change the device parametrics. Such a change, for example, could be a loss in speed of the device due to a process marginality or high resistances. Table 3.1 looks at the issue of yield loss from these two types of failure modes. The examples are for illustration only and this is not a complete listing of each category. Classification of yield killers is not always easy; at times the cause may be systematic and at other times random in nature. Particles are a type of defect that may fall in either systematic or random categories depending on the source of the problem and downstream processing.

### 3.4.3 Yield enhancement

Upfront investment of time and effort to understand and quantify the yield impact due to design, process, and manufacturing techniques helps in the long run in achieving profitability. Short-term fire fighting and band-aid solutions may solve the immediate yield problems, but these problems may remain a nagging source of low yield over the long haul. Let us look at the single most important aspect of each of the three components to wafer fabrication—design, process, and manufacturing. Table 3.2 highlights some of the most important aspects for each of the three components.

**Defect-tolerant circuitry.** As we move to higher and higher levels of device integration, there is demand for designs with a high degree of defect-tolerant architecture. Memory chips have moved in this direc-

**TABLE 3.2 Key Thrusts for Achieving High Yields**

| Area of yield improvement | Key modulator | Direction of change |
|---|---|---|
| Design | Defect-tolerant circuitry | Enhance defect tolerance |
| Process | Variation | Reduce process variation |
| Manufacturing | Cycle time | Decrease cycle time<br>Increase cycles of learning |

tion by incorporating redundant circuits; this has been done at the expense of some Si real estate. Should a particular bit, row, or column fail due to a defect, additional columns or rows may be activated to compensate for the one that is lost. For example, this may be done by using fusible links in the circuit. Reconfiguration may be also done through the use of lasers for altering the wiring layout.

Nonmemory architectures are constrained to reroute internal communication paths around defects to make them more defect-tolerant. Several defect-tolerant designs have been proposed that go into greater detail on this subject. To design circuits that are defect-tolerant, a good understanding of the defect and yield models is necessary. The relationship between the yield/defect modeling and defect-tolerant designing is highly complementary.

**Process variation.** In the case of processing, the single most important thrust is in reducing the variation for every process, once it has been developed and is ready to be used for production. To be used for production, a process must be deemed to be repeatable, predictable, and controllable. Continuous improvement is a key to maintaining and improving yields. The reduction of process variation needs to happen across the board for all the process steps. Let us take an example from the contact formation step and try to understand the effect of the variability of the processes on the final result—contact resistance. The contact formation step, as described in Chapter 2, is influenced by the field oxide, polysilicon, and BPSG thickness in a typical complementary metal-oxide semiconductor (CMOS) process. The variation in each of these processes has a cumulative effect on the overall thickness of the step. In worst-case situations, the contact may be formed with a depth equivalent to this step. The contact etch process itself is influenced by its own variability, namely, the etch rate. It is possible for wafers to experience a worst-case scenario of film thickness and etch rate. In order to compensate for such an event, the etch process contains an overetch component of time to ensure that all contacts are opened. This cumulative effect is part of the reason for the variability in contact resistance. If one considers the multitude of interconnect steps, the effect of process variation is significant. This makes yield forecasting all the more difficult and unpredictable.

**Cycle time.** The issues concerning cycle time were discussed in Section 3.2.6.

### 3.4.4 Yield models

How can we predict the yield of a particular device? What should be the yield target for each quarter? These are just a sample set of questions that must be asked in order to improve the yield for a process and to exceed the targets. Note that the yield models can not forecast the yield but help determine the current status and set realistic goals for different devices. Most yield models are based on random defects impacting yields.

**Total yield.** Systematic and random defects together determine the overall yield of the process. The systematic yield loss is not dependent on critical area, unlike the random yield loss component. Critical area may be considered as that area of circuitry that is susceptible to failure due to defects. Therefore, a first-order yield model consists of two components, one dependent on the die area and the other not. This may be expressed as

$$Y = Y_s \, Y_r \, (A_d, D) \tag{3.3}$$

where $Y$ = total yield
$Y_s$ = yield due to systematic defects
$Y_r$ = yield due to random defects
$A_d$ = area of the die
$D$ = defect density

In any process flow, critical masking steps may easily be identified if they affect the defect density contribution. Such a classification helps in determining the random defect density for a select number of layers instead of all the layers in a process flow. Once the critical masking steps have been identified, the number of critical operations may also be identified that are part of the critical layer.

If $n$ is the number of critical operations in a process flow, then the yield may be expressed as

$$Y = Y_s \sum_{i=1}^{n} Y_{ri} \, (A_{di}, D_i) \tag{3.4}$$

For accurate yield modeling, both the systematic and random yield components need to be modeled. There are several random yield models and the yield statistics associated with each are described extensively in the literature.[7,8,12,13] All the models agree with one another when the yields are high and the die area is small. But with ever-increasing die sizes, the yield models and associated statistics need to be modified and a model must be chosen that best fits the empirical data.

**Random yield models.** A key component to the yield models is the magnitude and distribution of the defects. The models described below address one or more of the following defect distributions:

1. Uniform distribution
2. Nonuniform distribution
3. Radial distribution

In all the models discussed here, $Y_r$ is the yield due to random defects, $D$ is the defect density, and $A$ is the critical area.

**Uniform defect distribution.** The Poisson model, given by Equation (3.5) is well suited when the defect density is uniformly distributed across the wafer. In real life this is not the case, but within a small critical area the defect density may be considered as uniform. Under this condition, the yield obeys the Poisson equation

$$Y_r = e^{-DA} \tag{3.5}$$

That is, the yield will exponentially decrease if the defect density, the area, or both increase.

**Nonuniform defect distribution.** An important enhancement to the above model was based on the assumption that the defect distribution is not uniform, even though the defect density, $D$, is constant. Let $f(D)$ be a function that describes the defect distribution. As the model takes into consideration varying defect distributions, the yield may be expressed as

$$Y = \int_0^\infty e^{(-DA)} f(D)\, dD \tag{3.6}$$

where $f(D)$ is the probability density function

$$\int_0^\infty f(D)\, dD = 1 \tag{3.7}$$

The above yield equation was developed further and several distributions were evaluated under various conditions. It has been shown that the triangular probability density function best describes the yield that is a function of a nonuniform defect distribution. Therefore, the yield model that is given by

$$Y_r = [(1 - e^{-DA})/DA]^2 \tag{3.8}$$

The Seeds model, given by Equation (3.9), also assumes that the defect density is nonuniform across the wafer and from wafer to wafer. Here the probability density function is exponential and is best expressed by the following equation:

$$Y_r = e^{-\sqrt{DA}} \tag{3.9}$$

**Radial defect distribution.** Several defects show a radial pattern in their distributions due to some processing steps. For example, particle distribution in a spin rinse and dry system may exhibit a radial pattern as the wafers are rotated around the axis of the boat at high speed. In other words, the defect density is different at the center as compared to the edge of the wafer due to defect clustering. In view of this, some yield models take into consideration this radial variation in defect density.

*Example.* Let us use one of these yield models to determine the defect density (due to particles) at some critical backend step. How do we determine the particle density requirement for the process that we plan to use, for each critical module in the flow? The answers to these questions will help determine the target goals for particle reduction once the systematic yield killers are identified and eliminated.

A simple step-by-step methodology for determining the particle density requirements for a process is shown in Figure 3.17. This is a four-step solution to the problem and may be extended to other similar problems. One of the requirements for any yield model is to be able to breakup the yield for each critical masking step. Such a break up in yield by masking steps enables a focused yield enhancement as we can then determine the defect budget for each critical layer.

*Step 1: Identify critical process steps that can cause yield degradation due to particles.* In the transistor formation portion of a typical CMOS device, there are three critical steps:

1. Isolation area definition
2. Active area definition
3. Gate definition

For a two-level metal system the following are the critical steps:

1. Contact definition

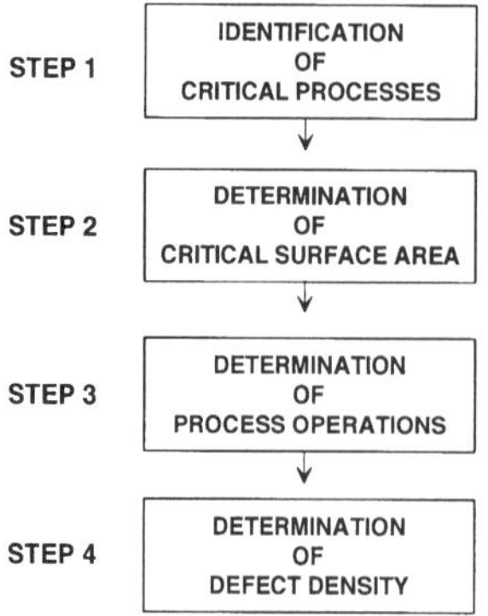


**Figure 3.17** Steps for determining particle density requirements.

2. First-level interconnect definition
3. Via definition
4. Second-level interconnect definition

The number of critical steps will increase as we have more levels of interconnects.

*Step 2: Determine the critical surface area for each of the critical steps.* Calculation of the surface area is important for understanding how much area may be effected by particle contamination. For the interconnect layers, the area is the product of the critical spacing and the length of the interconnect with this spacing. For the contact and via layers, the area is total contact and via area, respectively.

*Step 3: Determine the process operations that influence each critical step.* This helps to further break up the critical step into individual process step components that may influence the defect density. For example, the interconnect level may be split into three components consisting of the metal deposition, patterning, and etching. If the interconnect is a stack of several metals, the analysis may be modified to reflect this scheme.

*Step 4: Determine the defect density.* One of the yield models may be used for determining the defect density based on the empirical experience with the yield model, process maturity, and the yield improvement progression. Let the number of the process step components as defined in step 3 be $k$. Then the overall yield, $Y$, may be defined as

$$Y = \sum Y_k \tag{3.10}$$

Using one of the yield models, the defect density may be calculated for various levels of yields. As the process matures and higher and higher yields are realized, the appropriate defect density requirements for each critical process may be used as goals for the defect reduction activity.

Using the defect density number calculated from the above method, it is simple to determine the yield loss by equipment by using the above method. The focus of defect reduction may then be to realize a quicker gain in yield. Note that the above step-by-step method can become elaborate depending on the assumptions used for defect density determination.

### 3.4.5 Yield vehicle

A yield vehicle is often used to characterize and modify the process as necessary to achieve higher yields. In order to evaluate all the key electrical output parameters, a test product is used. This must contain test structures that can monitor parametric and catastrophic yield loss. For parametric yield loss, we can use the test structures, and SRAM chips may be used to monitor the catastrophic yield loss. The test and SRAM chips both may be printed on the wafer in a number of ways; one way is an alternating array of test and SRAM dice.

**SRAM.** The use SRAM chips as yield vehicles in combination with test chips has proven to be a viable yield monitoring vehicle. Such a vehicle is necessary, especially when the process is being developed for logic or custom IC applications. The SRAM has found a wide application as a yield vehicle because of several advantages that are illustrated in Figure 3.18. Owing to the repetitive pattern of SRAM circuitry, failure analysis is simplified. The failure analysis is accomplished in two ways. First, bit mapping can be used to isolate the failing rows and columns. Identification of the failing bits, rows, or columns helps identify the failing location for a thorough failure analysis by removing layer by layer, if necessary. If bit mapping capability were unavailable, then failure analysis would be time-consuming and would rely on a blind approach. Other memory devices may also be used as a yield vehicle.

As the intent of the SRAM chips are to flush most of the systematic problems, process development without such a chip is a definite handicap. Inline visual inspection for defects is much easier as the circuit pattern is similar across the die and defects are easy to locate. The defect location may be correlated to end of line failure analysis if the defect or its effect are still present. The SRAM chip, like any other yield vehicles, is only good for visual defects. Nonvisual defects, or rather their impact, may be analyzed by the various electrical test structures that should be used along with the SRAM test chip. Both of these together provide a powerful tool for debugging yield problems that may be either parametric or catastrophic in nature. Once the process is fairly well characterized, logic/custom devices may then be run with reasonable confidence that the yield loss on these devices is mostly due to random defects or design flaws. The SRAM yield vehicle establishes the baseline yield and defect density levels

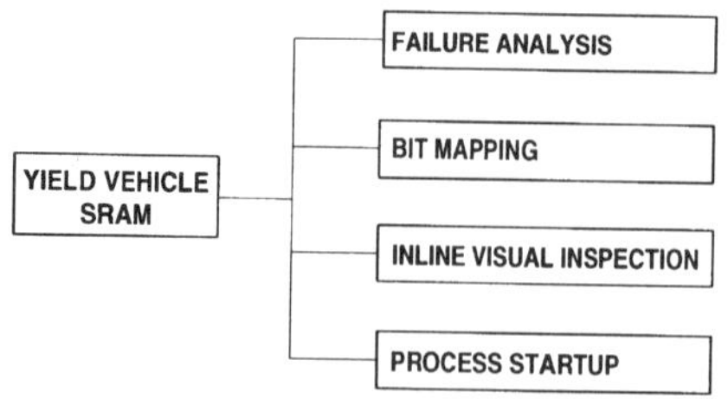


**Figure 3.18** Advantages of using SRAM devices as yield vehicles.

that other products may be compared against and improved on. Let us examine the test structures that complement the SRAM chip for yield analysis.

**Test structures.** Once the process flow has been defined, the next step is to develop a method of testing the various device parameters to make sure that the flow is basically sound. The primary function for the electrical testing of various structures is to identify processing problems and initiate fixes as appropriate. In addition to this, electrical test structures are used for yield enhancement and as reliability screens. Test structures are designed and placed in a specially designed test chips. The test structures are used for understanding transistor functionality, verifying design rules in Si, and monitoring the process (like contact resistance) both from a yield and reliability perspective. Due to Si real estate limitations, it may not always be possible to incorporate all different kinds of structures on the die. Here again, a compromise between information needed for process development and process sustaining needs to be made. Products that serve as yield vehicles may contain an elaborate matrix of structures that can provide information on various processing issues.

The electrical test structures may be divided into four distinct groups, as shown in Figure 3.19. This figure shows the various types of electrical structures that may be placed on a wafer. These four types or variations of these are aimed at detecting a cross section of process problems and defects. For multilevel interconnects the electrical test structures are not as many as required for the transistor formation. We can use structures from each of the above four categories to determine the various issues associated with multilevel metallization. The most common information that is pertinent to interconnects and that is extracted from the electrical test structures is shown in Table 3.3.

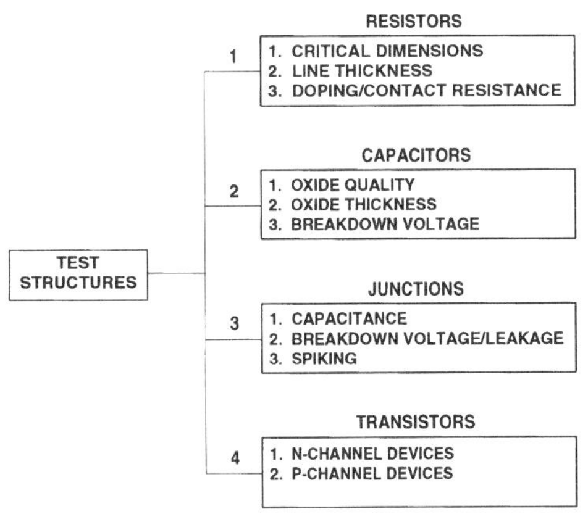


**Figure 3.19** Four basic device and process characterization test structures.

**TABLE 3.3 Backend Testing**

| Resistor | Capacitor | Junction | Transistor |
|---|---|---|---|
| Contact resistance | Dielectric breakdown and thickness | Leakage | Process-induced charging |
| Opens/shorts | | | |
| Misalignment | | | Threshold voltage |

Let us review the three resistor test structures (contact resistance, opens or shorts, and misalignment) and examine how the information may be used to monitor the process to gain an idea of the scope and nature of electrical testing of these structures.

**Contact resistance.** A detailed contact resistance cause and effect diagram was presented in Chapter 2. Contact resistance monitors are used extensively to debug and monitor the process. These resistance measurements are generally made by using Kelvin structures with four pads. Two of these pads are used to force the current and the other two are used to measure the voltage. To obtain an average interfacial resistance, the voltage measurements may be made by reversing the current through the original pads and by repeating the measurement with interchanged terminals.

To measure the effects of topography and processing conditions on contact resistance, contact chains are used. In case of CMOS devices, these contain several hundred to thousands of contacts over varying topography, silicided diffusion (n+/p+), and polysilicon. The contact and via chains are placed in the scribe streets or in test chips that are strategically located on the wafer. Figure 3.20 shows the top view of a contact chain consisting of contact links. Current is forced through the chain and the resulting voltage drop across the pad is measured.

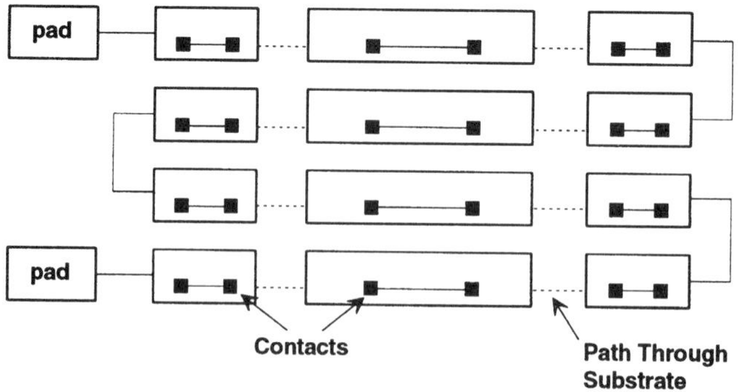


**Figure 3.20** A schematic top view of a contact chains test structure for determining contact resistance.

The chains are measured at different locations across the wafer to obtain within wafer variation of resistance. All the wafers in a lot may be tested, and the contact resistance trend by lot provides a very good picture of the lot-to-lot variability and any excursions. The variation of the contact resistance across a wafer may be used to check for systematic processing problems. Contact resistance failures usually have a strong correlation to yield and therefore are primarily used as yield monitors.

A cross-sectional view of these contact links is shown in Figure 3.21. This is a schematic representation of the contact links over diffusion. As shown by the arrows, the current, *i*, is forced from one end to the other. The current travels through the metal interconnect and into the plug (*P*; if used) that connects the diffusion or polysilicon to the interconnect. Due to the layout of this chain, the current is forced to travel through all the plugs, provided the contacts are not open. The number of contacts in a chain is dictated by the Si real estate that can be allowed for test structures. Obviously, with more chains the capture rate of processing defects will be higher.

**Metal shorts/opens.** As metallization pitch has decreased, the need to check the integrity between metal interconnects has become very evident. Figure 3.22 shows two generic structures, which may be designed with several variations, for testing for metal shorts or opens

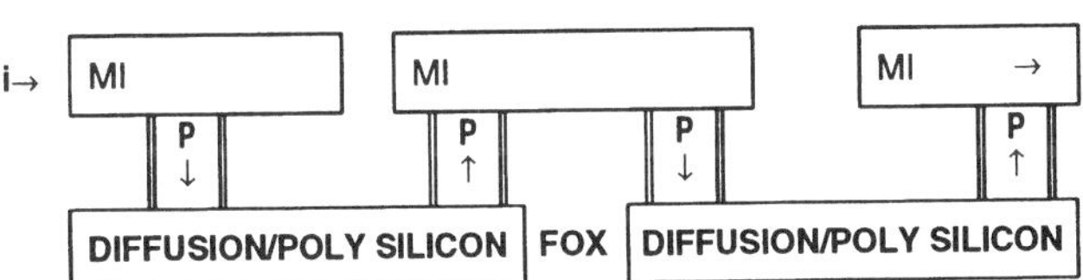


**Figure 3.21** A schematic cross-sectional view of contact test structures.

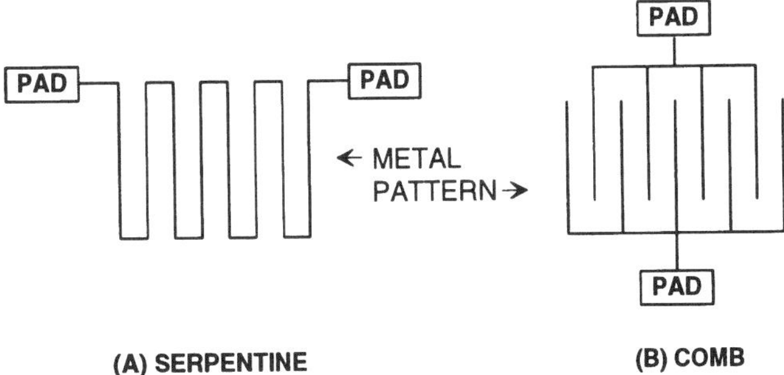


**Figure 3.22** Metal serpentine and comb structures for determining shorts and opens in conducting layers.

at the same level or between levels of interconnects. These shorts or opens may be caused by various processing defects, for example, Al hillocks. To capture the various conditions, both the serpentine and comb structures are defined over varying topography. The electrical measurement is once again a resistance measurement, and the current is forced through the structure and the voltage drop across it is measured. These structures may be used for electromigration evaluation also.

This concept may be extended to monitoring the defect density of a piece of equipment or a specific segment of the process. Wafers with dice having serpentine and comb patterns may be defined and processed through a small segment of the process that is under evaluation. Testing these structures for opens and shorts will provide a clear picture of the defects that cause shorts or opens and therefore can be traced to particular equipment. Processing such wafers at regular intervals through various segments of the process will help in determining the electrical defect density baseline. The impact of any process changes may be quickly checked by processing such wafers and the data compared with historic values. Furthermore, these wafers may be visually inspected and the defect that caused an open or short characterized.

These structures are useful for obtaining information on a host of issues. This information may be obtained within days as the process flow for generating these structures is simple and consists of only a few steps, as shown in Figure 3.23.

**Misalignment.** The registration between layers, especially critical layers, is an important process control and process development tool. Misalignment may be measured electrically or by using optical means at each lithography step. Electrical test structures can be used to monitor the degree of misalignment and its impact on process problems.

To measure the degree of misalignment between layers using an electrical test structure, a simple resistor with multiple taps for voltage measurement, may be designed to detect the misalignment in

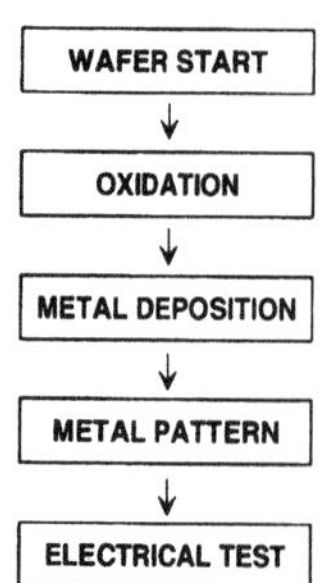


**Figure 3.23** A simple process flow for generating defect monitors with serpentine and comb structures.

both the $x$ and $y$ directions. When there is no misalignment the voltage difference between these taps will be zero. With misalignment the resistance changes and the difference between the taps may be measured. This difference gives both the magnitude and direction of the misalignment. This result is an indication of how well the lithography process is controlled. As an aid in verification, optical and visual misalignment structures may be placed next to the electrical structures.

The structures shown in Figure 3.24*a* and *b* and their respective cross sections *c* and *d* are designed to understand the implications of misalignment on the circuit functionality. These structures show misalignment of the metal interconnect (MI) over a contact or via plug, if used. In an ideal situation, the metal would completely cover the plug; in a misaligned state, part of the plug is exposed. Structures with various degrees of misalignment may be designed and placed on the wafer to study the impact of misalignment on contact resistance. For example, contact resistance structures may be defined with misalignment in both the negative and positive directions with varying magnitudes of misalignment. If contact resistance increases as a result of misalignment, one of the reasons may be plug damage. This damage is a real problem as the plugs would be exposed to the metal etch and cleans. During the metal etch process, the plug may be partially etched or the interface between the metal and plug may be contaminated. The robustness of any process depends on its ability to tolerate some amount of misalignment. The design rules should allow this as long as there is no detrimental impact on the electrical parameters.

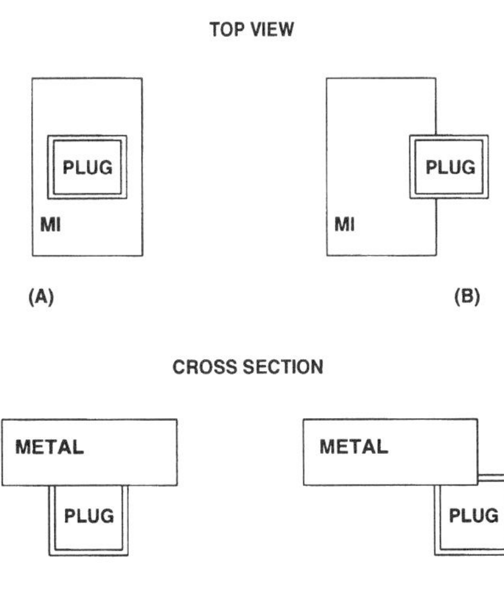


**Figure 3.24** Pattern misalignment test structures showing top and cross-sectional views.

## 3.5 Defect Reduction

Achieving zero defects, and therefore high yields, is definitely the goal, but the cost of doing this should also be evaluated. Let us discuss this subject by breaking it up into four key defect reduction[24–39] groups as shown in Figure 3.25 and examining each one in the context of multilevel interconnect processing.

### 3.5.1 Defect identification

Every process will be affected by defects that may depress yields. The key to success is identifying these before significant and severe damage is done to products. Defects should be identified inline, that is on a real time basis, as wafers are moving through the production line. As Figure 3.26 shows, defect detection may be done using visual inspections at various locations in the process flow. In addition to this, particle (a loose definition is used here to cover all defects) counters may be used to determine size and location of defects. Lastly, inline electrical testing of serpentine and comb structures provides valuable information on defect levels. If possible, the electrical defect must be correlated to the visual defect in order to understand the failure mechanism.

Offline (end-of-line) defect identification is more elaborate as several failure analysis techniques are used that may not be used in the wafer fabrication clean room, for example, gold coating of samples for scanning electron microscope (SEM) analysis. Offline defect identification involves isolating the failing bits, rows, and columns using bit

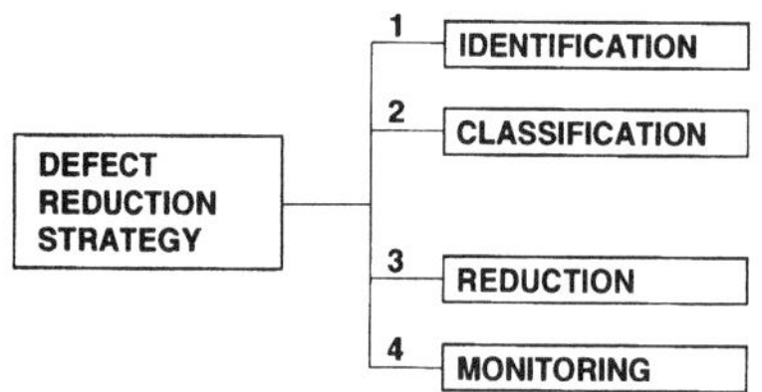


**Figure 3.25** Four major areas of focus for defect reduction.

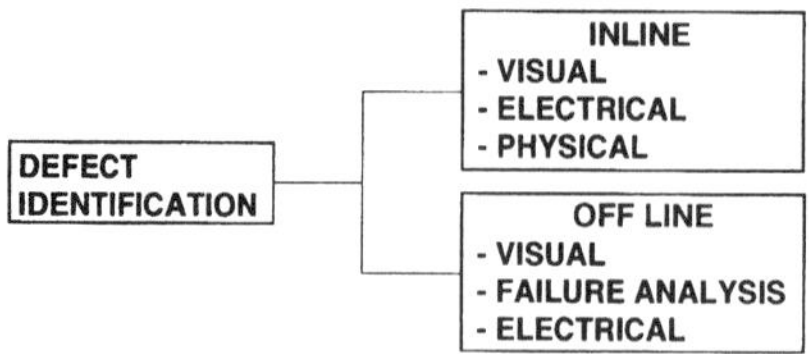


**Figure 3.26** Inline and offline defect identification methods.

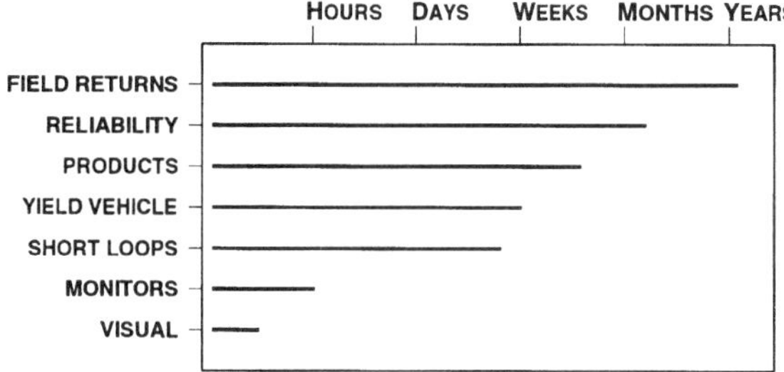


**Figure 3.27** Comparison of the delay in information feedback from several defect identification/monitoring methods.

mapping and carefully removing layer by layer until the defect that has caused the failure has been found. Once the defect has been found, various visual images are taken. If required, an elemental analysis is done to learn more about the nature of the defect. Obviously, this is a time-consuming task. It is important at this juncture to examine the various defect identification methods and their throughput time for analysis. Figure 3.27 compares the various defect identification methods (listed on the y-axis) with the throughput time (x-axis). The throughput times given here are simply for illustration purposes and are not absolute times. The throughput time would depend on the device type, processing cycle time, etc. From this we can conclude that the greatest leverage is in inline defect monitors.

**Field returns.** An indication of the interaction of a defect and the environment over a long period of time is determined by the failure rate of the devices out in the field. Failure analysis of these failing parts can help in determining the failure mechanism. Feedback to the factory may be late and the defect that caused the failure may or may not still be present. However, the defect identification of field returned parts may shed light on any systematic or random problem with the process. For example, corrosion of interconnects usually gets worse over time and this may cause a failure. If this is a systematic problem, it is then appropriate to go back and revisit the interconnect definition process.

**Reliability.** For a faster look into the quality of the contacts, interconnects, and dielectrics, reliability tests such as burn-in, steam test, and electromigration are used to determine the failure rate of the parts. The data from such reliability tests takes time as devices have to be packaged and the tests performed at different conditions to simulate real-life conditions. These tests can not be used as real-time defect reduction methods, but can provide baseline data for comparison as material is sampled for qualification of a new process or a major process change.

**Products.** On a more frequent basis, probe yields give a reasonable picture of the level of the defect density over time since yield is exponentially inversely proportional to defect density. To determine the defect density baseline, wafers with a certain yield can be selected at regular intervals and defects determined using offline methods. In this manner, the baseline defect density for each layer can be determined as a function of time.

**Yield vehicle.** For a more proactive method for defect reduction, two methods may be used. The first is to map out electrical defect density and the other visual defect density. In some cases these two correlate and in others they may not. The reason for this is that some defects may not degrade the yield.

**Short loops.** To obtain the electrical defect density of a section of the process or just a few steps, various short loops may be used. These can have regular test structures for measuring the resistances, functionality, or just structures that have been designed for measuring the density of opens or shorts. These short loop process flows are powerful aids in establishing the defect density baseline or for the qualification of new equipment at a process step. The throughput time for these short loops may range from a few days to a couple of weeks.

**Monitors.** For visual defect characterization, SEMs or inline automated defect inspection tools may be used. These tools could be particle counters or more elaborate tools that can count and classify defects through several layers. The feedback is immediate on the type of defects and, if the problem is significant, the production line may be stopped and the defect mode investigated. Inline visual inspection is well positioned to catch excursions and prevent good material from being processed through suspect process steps. The last monitor that helps present a defect density picture is the equipment particle monitor. These monitors highlight any excursions in equipment that might contaminate product.

### 3.5.2 Defect classification

The defects identified are classified according to their size and nature. As feature sizes decrease, smaller and smaller defects impact the yield and therefore it is important to classify defects by size. The nature (size, shape, elemental makeup, etc.) is important as it helps focus on a process module that could be the source of the problem. Classification of defects by module and by defect types is essential to get an overall view of the defect scenario. New defects are classified as they are discovered and added to the defect library.

**Defect density presentation.** Defect paretos are an excellent graphical tool to observe the progress of the defect reduction effort and also to direct valuable resources to addressing the highest leverage defects. In addition to pareto charts, defect status may be presented as shown in Figure 3.28.

In this chart, the defect density for each, is presented and gives an overall picture of the defect levels in the factory. The defect density numbers do not pertain to any particular factory and are just used for illustration purposes. The total defect density may be used to determine the yield from the yield models. This gives a clear signal for the defect reduction activities to focus on.

**Defect types.** The discussion so far has focused on providing an overview of the manufacturing, process integration, yield issues, and techniques for monitoring the defects. This section presents some of the common defects affecting multilevel interconnects that degrade yield and reliability of ICs. Integrated circuits with multiple levels of metallization have a greater propensity for defects as they entail more processing than an integrated circuit with just a single level of metallization. To aid the reader in understanding defects and how they impact the process and/or product, each defect is discussed as follows:

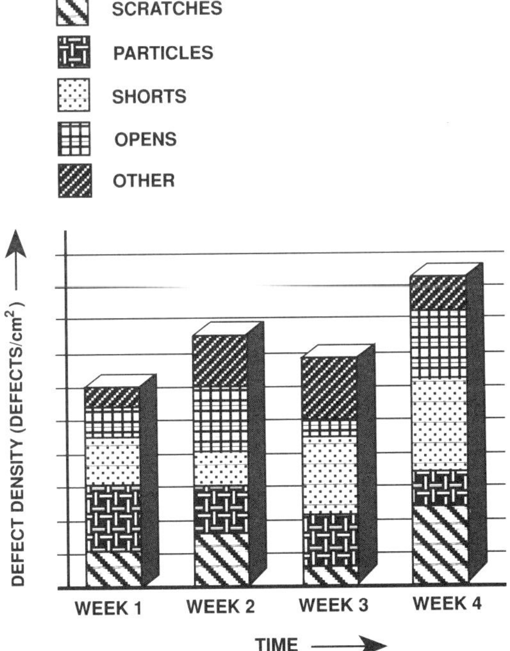


**Figure 3.28** A histogram of defect density variation over time.

1. Defect source
2. Impact on yield and reliability
3. Process fix

Such an analysis helps in the evaluation, elimination, and prevention of the defects. The origin of defects may vary but the effect of may be classified into two main defect categories. The two categories are as follows: shorts and opens.

**Shorts.** These are defects that are electrically active and alter the current path when the defects bridge two interconnect lines that are physically separated by an insulator. Figure 3.29 shows a typical example of a short between two conducting paths.

**Opens.** These are defects that cause a discontinuity in the conduction path. This discontinuity may be missing interconnect material or an interaction with other defects that might have prevented the formation of the interconnect path. Figure 3.30 shows opens in interconnect lines due to defects.

The following discussion considers a few typical defects that affect multilevel interconnect processing. These are divided into two categories: systematic and random. It is not always easy to classify the defects in such a manner. Defects may switch from one category to the other easily depending on the nature of the defect source, which may be the process or equipment. Table 3.4 highlights some of the common defects.

Let us take an example of this dilemma in defect classification. If a particle were to create a defect on the mask then this defect occurrence may be considered as random in nature. However, if this mask is then

**Figure 3.29** An example of shorted conducting paths.

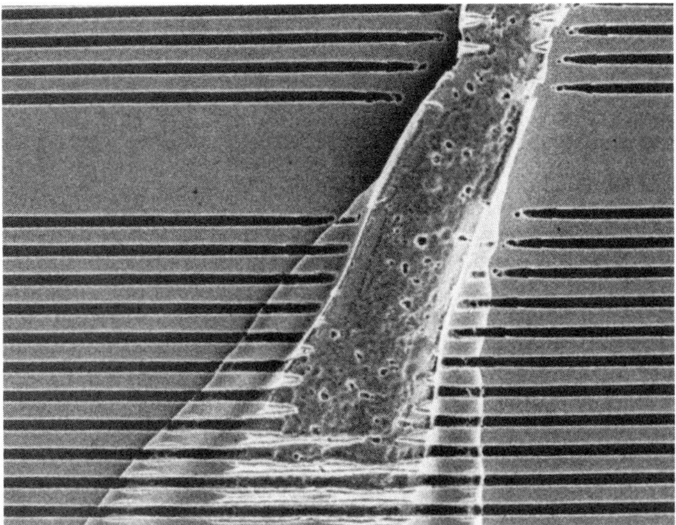

**Figure 3.30** An example of discontinuity (opens) in a conductor.

**TABLE 3.4 Common Defects**

| Systematic | Random |
|---|---|
| Corrosion | Particles |
| Hillocks | Mask defects |
| Edge effects | Scratches |

used to print the pattern on the wafer, we would have a systematic failing location owing to the step and repeat operation of the stepper.

### Corrosion

**Source.** The primary source of corrosion is the metal etch process chemistry. Aluminum interconnects are etched using chlorine (Cl) which is readily absorbed in both resist and the oxide surface. Corrosion is observed if the wafers are exposed to atmosphere following etch without passivation or some chlorine removal step beforehand.

**Impact.** If corrosion is widespread, it may be possible to detect it on wafers while they are still in the clean room. Sometimes, corrosion of metal interconnects takes a long time and could cause a failure after the parts have been packaged and shipped. Corrosion may either create shorts or opens. Shorts are created when the corrosion byproducts connect two adjacent interconnects. In some cases, corrosion removes part of the interconnect material, leaving a discontinuous or open interconnect. Figure 3.31 shows two possible ways corrosion can

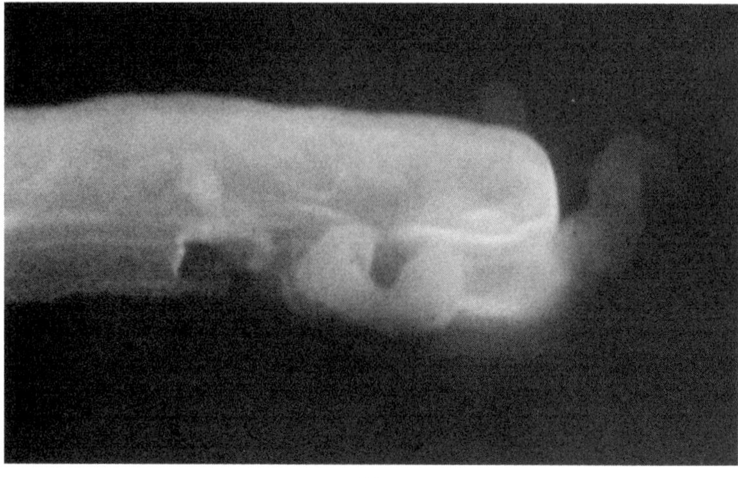

(a)

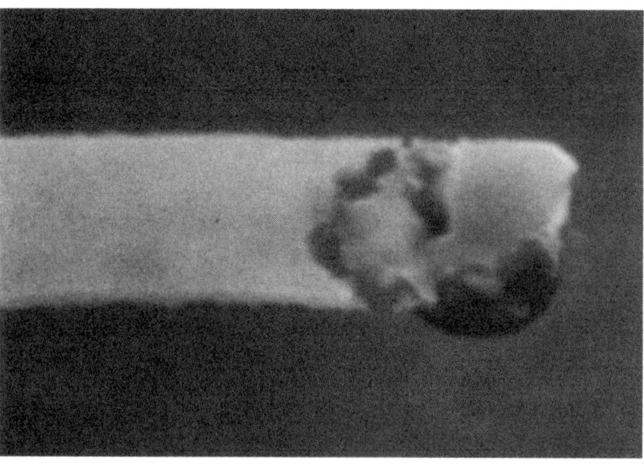

(b)

**Figure 3.31** Corrosion in metal interconnects. (*a*) Creating a potential short. (*b*) Causing an open.

cause device failure. Figure 3.31*a* causes an interconnect short and Figure 3.31*b* causes an interconnect open.

The impact of corrosion may be observed in several ways depending on the severity of the problem. Parametric testing after wafers have been completely processed may show a change in metal interconnect critical dimensions. In addition, shorts or opens may be observed between interconnects. These problems translate into yield loss if the corrosion reaches an advanced stage.

**Fix.** Corrosion problems associated with metal etch and cleans may be solved by an effective in situ passivation step that is a part of the etch process and is followed by an efficient resist cleaning process. Immediate substitution of Cl by fluorine (F) absorbed in the resist followed by resist stripping are done to prevent corrosion. Capping the Al interconnect with a dielectric seals the film and prevents further corrosion.

### Al hillocks

**Source.** One of the common problems with Al interconnects is the formation of hillocks. *Hillocks* are extrusions of Al from the surface of the film that are caused by the compressive stress in the metal film. The problem is severe for Al interconnects because of the thermal mismatch between Al and Si, especially since Al adheres very well to Si and $SiO_2$. When the Si wafer is heated, the different rates of thermal expansion cause the Al film to expand more than Si or $SiO_2$. This resulting stress causes formation of hillocks. Compressive stresses in the Al film become high as the temperature exceeds 300°C—a very easy temperature limit to exceed in most chemical vapor deposition (CVD) dielectric film deposition processes. In addition to this mechanism, hillocks may also be formed by the high rate vacancy diffusion in the Al films. Figure 3.32 shows a typical Al hillock that has been formed after thermal treatment.

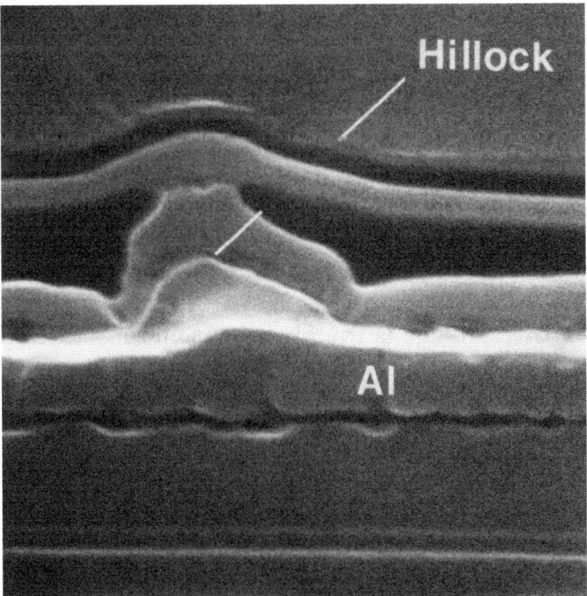


**Figure 3.32** A hillock in an aluminum film.

**Impact.** The impact of hillocks on interconnects is detrimental to the device functionality. Hillocks cause shorts in both the lateral and vertical directions. In the lateral direction, two adjacent metal lines may be shorted—a problem that gets worse as the metal pitches are reduced. In the vertical direction, the hillock is covered by dielectric. In subsequent via definition steps the resist over the hillock will be thinner than in other locations. During the etch step, the resist erodes quickly, exposing the dielectric to the etch, which may completely remove it. The hillock could short to the next level of metallization.

**Fix.** There is no one comprehensive fix for the hillock formation problem. There are, however, several techniques[5] that help reduce the hillock formation. A few of the key ones are given here. Any fix to the hillock problem must be considered along with the other integration issues associated with changing the composition of the metal stack.

1. Restricting the dielectric deposition temperature to around 350°C. In addition, the hillock size and density may also be minimized if a low compressive stress dielectric film is deposited over the metal quickly enough.
2. Capping the Al film with a layer W or Ti.
3. Adding Cu to the Al film. This has been shown to minimize hillock formation as the Cu precipitates in the grain boundaries and prevents vacancy migration.

### Edge effects

**Source.** The wafer edge is a source of defects which may not stay at the edge but may cause shorts/opens anywhere on the wafer. These defects may be transported and deposited on the dice as the wafers go through various processing steps. Figure 3.33 shows the wafer edge that has been subjected to the extremes of various processes. Particles and other defects around the edge could easily cause problems. Figure 3.34 shows the area on a wafer that is susceptible to damage.

One of the sources of this damage is the placement and interaction of the various machine clamps that are used to hold the wafer in place. Damage can also occur when resist is removed from the wafer edge as part of the photolithography process. This edge bead removal process, as it is called, may intrude in from the wafer edge. Since dice are not printed in the shaded area, microloading effects during the etch process further damage this area due to no pattern and inadequate film thickness, arising from film non-uniformities at the edge.

**Impact.** The impact of not controlling and monitoring edge effects can be severe as edge dice adjacent to the shaded area may not be

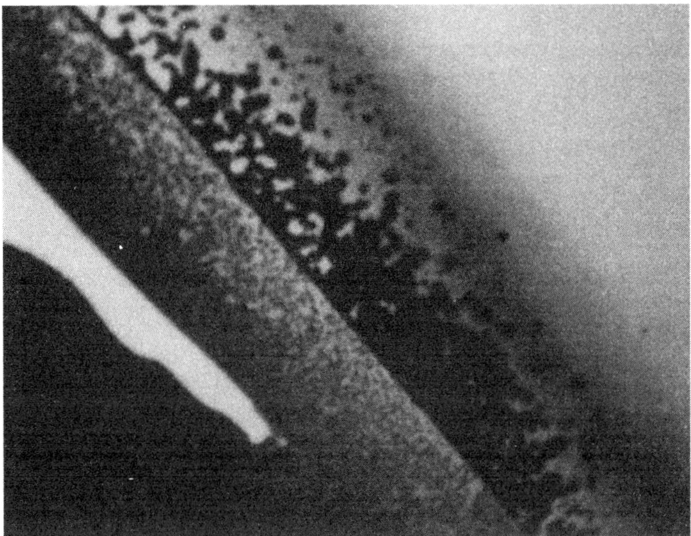

**Figure 3.33** Wafer edge damage during processing highlighting a potential source for defect generation.

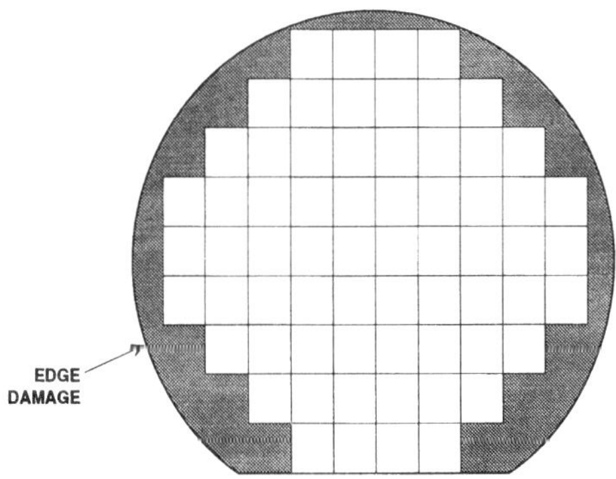


**Figure 3.34** A wafer map indicating areas of damage around edge.

reliable due to defects from the wafer edge. The short is the common failure mode associated with edge contamination. This translates into yield loss not only on the edges but sometimes in other locations.

**Fix.** Can this shaded area be reclaimed and die printed there? Perhaps some amount of Si real estate may be recovered with optimization of the deposition, etch equipment setup, and other processes. We can not help but lose a fraction of the area around the wafer edge. The short-term fix in most cases is to compensate for aggressive etches around the edges with a thicker oxide film. The long-term fix is a combination of equipment and process modification. One other technique is to print all the dice to the very edge.

### Particles

**Source.** Particles may or may not cause yield loss, but it is good policy to reduce the particles in the production area and on the wafers. Figure 3.35 is a cause and effect diagram for particle generation that is essentially an expansion of the branch given in Figure 3.16. Particles may be deposited on the wafer surface due to the mechanical movement of wafers through equipment process chambers. The wafer transport mechanisms in most machines are a major source of particle generation. What about process induced particles? This is the other main component to the problem of particulate contamination. Contamination can be generated from any process steps. For example, process induced particulate contamination in dielectric or metal etch chambers is a major concern. The etch process relies on polymer formation to control profiles. This polymer gets deposited on chamber walls and flakes onto the wafers. Figure 3.36 shows a particle that has been deposited from chamber walls in a deposition system.

**Impact.** Particles may be benign at the layer they were deposited, but may cause catastrophic yield loss at some downstream location. This occurs when the particle serves as a nucleation site for downstream depositions of metals and has a layer of metal deposited over it. After the etch process, this particle could have a metal spacer around it that could short interconnect lines. This is one mechanism where small particles increase in size as they are exposed to downstream processes. Particles degrade yield by causing shorts, opens, or both.

**Fix.** Obviously, it is of paramount importance to reduce the particulate contamination levels. One way is to improve the equipment that is used and the other is to develop processes that operate within a window of very low particulate contamination. Equipment-related particle generation is one area for continuous improvement to realize defect density goals for each part or module of the process flow. Control and reduction of particles in process equipment is all the more critical when the same set of machines is used several times in the process. For example, metal deposition machines are used several times for interconnect definition as described above. In some cases, reduction in particles is only possible through the use of wet cleaning procedures. Unfortunately, wet cleaning options are limited for interconnect formation steps. Wet cleans may be divided into two modes of application.

1. *Film removal.* This is usually removal of photoresist and some amount of dielectric. Incomplete removal of these films may cause catastrophic failures. For example, after metal etch, the photore-

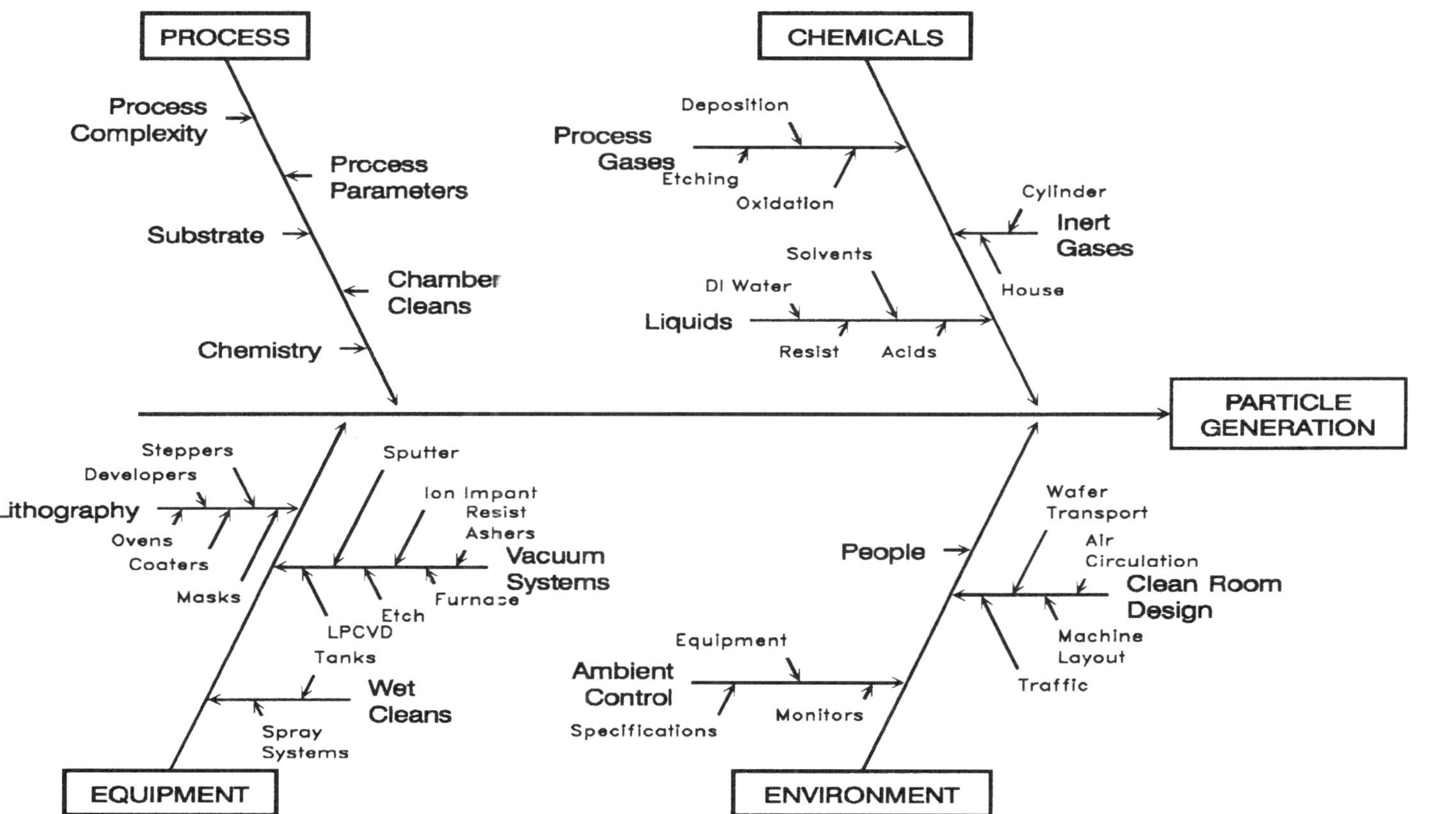


**Figure 3.35** Cause and effect diagram for particle generation.

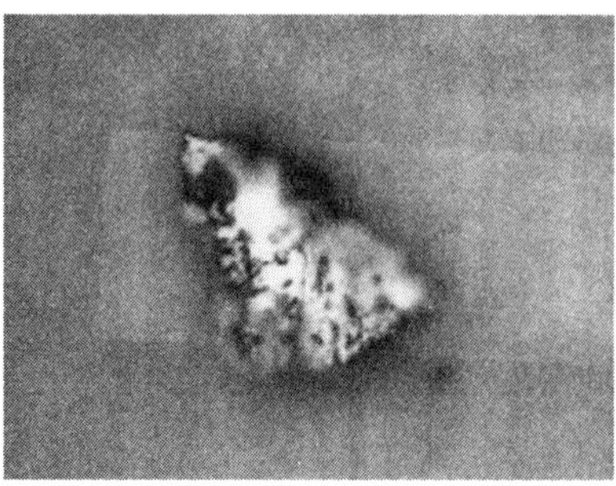

**Figure 3.36** An example of particle generation from thin-film deposition systems.

sist must be removed and the Cl absorbed surface cleaned to prevent metal corrosion.

2. *Contamination removal.* This is critical for contact and via layers where residues, particles, and other contamination may prevent good contact and cause the resistance values to increase. Cleans used here must be compatible with the metal layers and environmental requirements.

#### Mask defects

**Source.** A reticle or mask defect may be random in nature and might have been caused by particles or by residues from the mask making operations. However, this defect becomes systematic in nature after the reticle is used to print die on Si. Let us assume that the reticle field consists of two dice that are stepped across the wafer. If one of these has a defect, then the defect is also printed on the wafer as the stepper moves across the wafer. Figure 3.37 shows the situation where we have a repeating bad die that alternates with a good die.

**Impact.** The impact of such a defect is catastrophic and results in a 50 percent yield loss, assuming that the reticle field consists of two dice. The defect could cause shorts or opens depending on its nature and placement.

**Fix.** The fix for this problem must start in the mask making operation. Stringent quality control of the products (reticles) is required. Once these reticles arrive at the fabrication site, careful inspection and sample printing across the wafer is often done to ensure that no pattern defects exist.

#### ILD pinholes

**Source.** The primary source of interlevel dielectric (ILD) pin holes is the thinning and subsequent stripping of resist over the tallest parts of the die during contact/via etching. Aluminum hillocks and particles also cause pinholes in the dielectric film.

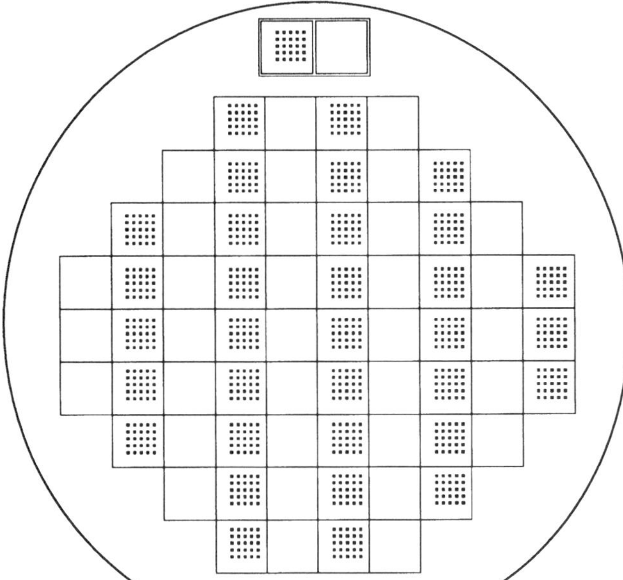

**Figure 3.37** Yield loss due to repetitive mask defects.

**Impact.** Location of the pinhole determines the impact on circuit functionality. If the pinhole were to fall over an area where two conductors overlap, then these would be shorted as there is now a path for conduction between them. In such a case the yield is degraded. Reliability of the die will be degraded if the dielectric has pinholes. These pinholes can trap moisture and other contaminants. This is especially true for the passivation layer which is deposited to provide a hermetic barrier and protect the circuitry underneath. Pinholes in the nitride passivation layer can cause reliability failures. A quick check for pinhole density in the dielectric film is to dip the wafer in an acidic solution. If pinholes are present, then the acid will attack the Al film underneath as it penetrates through the pinholes.

**Fix.** Pinholes may be prevented by reducing the particle density and hillock formation. Increasing resist thickness and improving the oxide to resist selectivity are some of the techniques that may prevent pinholes from forming. Of course, the thin oxide film that is deposited must be of good quality to begin with. In some cases multiple depositions are done to reduce pinhole occurrence.

### Scratches

**Source.** The source of scratches is primarily wafer handling within equipment itself or by people handling wafers. Scratches may be a random defect but this is really a systematic problem that impacts all processes.

**Impact.** Scratches may rip the interconnect lines and cause opens. The scratches will be areas of high defect density as the wafer moves through the process. There is always a danger that defects from these scratches may contaminate neighboring dice.

**Fix.** No or minimal wafer handling by manufacturing operators is the ultimate solution. Equipment-induced scratches may be addressed with modifications to the wafer handling system. Automated cassette to cassette inspection and monitor systems prevent the need to handle wafers. Automated systems for wafer transfer are used to minimize handling.

### 3.5.3 Defect reduction

Having identified and classified the defects, the next major program is the elimination of these defects. From the yield loss and the particle cause and effect diagrams, we observed that there are several sources of defects. The defect contribution varies from factory to factory, quality of equipment, process, and raw materials. Figure 3.38 shows a pie chart with a possible breakup of the defect contributors to

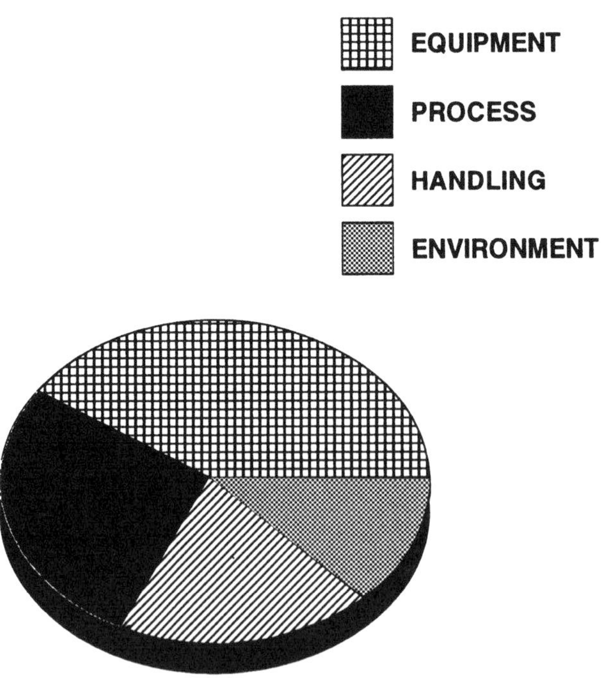


**Figure 3.38** Sources of defects.

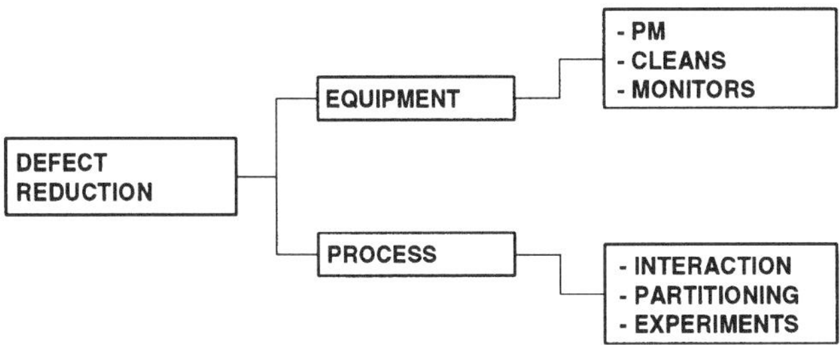


**Figure 3.39** Defect reduction directions.

overall defect levels. Such an illustration is useful in launching a factorywide campaign in defect reduction, involving vendors, improving incoming quality of materials, proper selection and evaluation of process equipment, ultimately enhancing the manufacturing methodology. If these basics are not addressed, defect reduction becomes that much more difficult. Let us examine the defect reduction thrusts for the equipment and process more closely, as shown in Figure 3.39.

**Equipment.** There is no question that the quality and design of equipment is extremely important in reducing defect levels. For example, a wafer movement mechanism based on belts and exposure to high temperatures tends to have higher levels of particles than a "pick and place" type of system. Improvements in equipment must be made by the equipment vendors to satisfy technology demands of the future.

Equipment downtime should be monitored and the causes addressed. Obviously downtime creates manufacturing bottlenecks. The solution is not in increasing the capacity by purchasing more machines, but in enhancing capacity by increasing the mean time to failure of the equipment. This is done by timely and thorough preventive maintenance (PM), even in the face of production pressures.

Some processes cause polymer or other chamber coatings on the equipment chamber walls. Flakes from the chamber walls may fall on the wafer surface, causing yield loss. Chamber cleans, either in situ or not, are needed to prevent an increase in defect levels. Equipment particle levels should be monitored on a regular basis using particle counters. Excursions in particle levels should be reason enough to shut down the machine, investigate the cause, and take corrective actions.

**Process.** Defects that become killer defects as a result of process interactions may be effectively eliminated by defect partitioning. Defect partitioning, a defect identification technique, involves inspecting the same area on the same wafer as it moves through the process flow. The results are studied by overlaying the location of defects at each step. This provides information on how each process changes the nature of the defect and contributes new defects. The success of defect partitioning depends on the accuracy of inspection tools, in locating defects and their coordinates. Pictures may be taken after each step so the changing appearance of the defects may be observed. Using detailed visual pictures and elemental analysis, a cause and effect diagram should be generated to highlight all possible causes of the defect(s). The various models that are generated by these problem solving exercises may be evaluated by split-lot experimentation. To speed up resolution of problems, people who have a broad and good understanding of the process should be consulted.

### 3.5.4 Defect monitoring

Defect monitoring procedures must be carefully designed and implemented since they tell us if we have been successful in defect reduction or not. This step may be divided into two activities, as shown in Figure 3.40. Earlier we discussed the techniques for establishing the baseline defect level for a process. The baseline defect level is presented either in trend or pareto charts and appropriate action is taken to continuously reduce the baseline. Once a defect excursion has been isolated and a fix instituted, it must be monitored to make sure that the defect is truly eliminated. Monitors for this purpose may be manual or automated inspection. Proper documentation is necessary to finally complete the whole problem solving process.

## 3.6 Summary

The focus of this chapter has been on manufacturing issues—merging the designs and processes to obtain high-yielding wafers. Yield enhancement is possible only when proper manufacturing practices are combined with defect reduction activities and cost boundary conditions are satisfied. Truly, the rate-limiting step in producing ICs in large volume ahead of the competition lies in the successful understanding of manufacturing technology. As process technology is pushing the level of integration to higher and higher levels, the need for more levels of metallization increases. Such an increase in interconnect levels leads to several cycle time improvements and defect reduction opportunities.

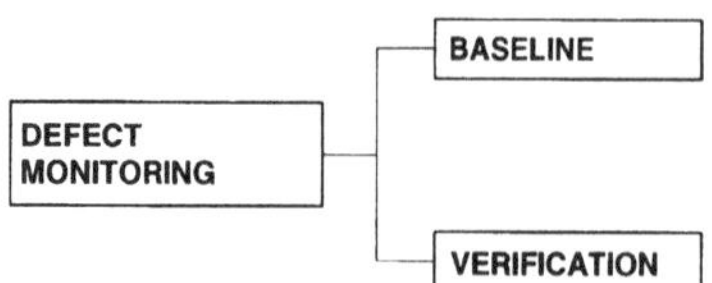


**Figure 3.40** Defect monitoring procedures.

## References


1. G. H. Bowers, "Continuous flow manufacturing," *Intl. Semiconductor Manufacturing Symp.*, 1990.
2. J. Cunningham, "Using the learning curve as a management tool," *IEEE Spectrum*, June 1990.
3. K. Suzaki, *The New Manufacturing Challenge*, The Free Press, 1987.
4. W. J. McClean (ed.), *Status 1992: A Report on the Integrated Circuits Industry*, Integrated Circuits Engineering Corp., 1992.
5. S. Wolf and R. N. Tauber, *Silicon Processing for the VLSI Era, Volume 1—Process Technology*, Lattice Press, 1986.
6. S. M. Sze (ed.), *VLSI Technology*, McGraw-Hill, 1986.
7. D. Moore and H. Walker, *Yield Simulation for Integrated Circuits*, Kluwer Academic Publishers, 1987.
8. W. Moore et al., "Yield modeling and defect tolerance in VLSI," *Intl. Workshop on Designing for Yield*, July 1987.
9. B. Hoyt, *The Implementation of Statistical Process Control (SPC)*, Northcon, 1987.
10. H. Parks, "Yield modelling from SRAM failure analysis," *Proc. IEEE Intl. Conf. on Microelectronic Test Structures*, Vol. 3, March 1990.
11. H. Parks et al., "Use SRAMs and test structures as yield monitors," *Semiconductor International*, April 1989.
12. C. Winter and W. Cook, "Interval estimates for yield modelling," *IEEE Journal of Solid State Circuits*, Vol. 4, August 1986.
13. C. Stapper et al., "Integrated circuit yield statistics," *Proc. of the IEEE*, Vol. 71, No. 4, April 1983.
14. C. Stapper et al., "Evolution and accomplishments of VLSI yield management at IBM," *IBM J. Res. Develop.*, Vol. 26, No. 5, 1982.
15. S. Dimitrijev et al., "Yield model for in-line integrated circuit production control," *Solid State Electronics*, Vol. 31, No. 5, 1988.
16. R. Lea and H. Bolouri, "Fault tolerance: Step towards WSI," *IEEE Proceedings*, Vol. 135, Pt. E, No. 6, 1988.
17. V. Flack, "Estimating variation in IC yield estimates," *IEEE J. Solid State Circuits*, Vol. SC-21, No. 2, 1986.
18. P. Feldman and S. Director, "Improved methods for IC yield and quality optimization using surface integrals," *Research Report No. CMUCAD-91-31*, Carnegie–Mellon University, 1991.
19. P. Feldman and S. Director, "A macromodeling based approach for efficient IC yield optimization," *Research Report No. CMUCAD-91-27*, Carnegie–Mellon University, 1991.
20. P. Feldman and S. Director, "Accurate and efficient evaluation of circuit yield and yield gradients," *Research Report No. CMUCAD-90-32*, Carnegie–Mellon University, 1990.
21. W. Maly and A. Strojwas, *SRC Technical Report T91118*, August 1991.
22. T. Michalka et al., "A discussion of yield modelling with defect clustering, circuit repair, and circuit redundancy," *IEEE Trans. on Semiconductor Manufacturing*, August 1990.
23. D. Dance and R. Jarvis, "Using yield models to accelerate the learning curve progress," *IEEE Trans. on Semiconductor Manufacturing*, February 1992.
24. R. P. Donovan (ed.), *Particle Control for Semiconductor Manufacturing*, Marcel Dekker, New York, 1990.
25. S. Strathman and S. Lotz, "Automated inspection as part of defect reduction program in an ASIC manufacturing environment," *SPIE*, Vol. 1261, *Integrated Circuit Metrology, Inspection and Process Control IV*, 1990.
26. V. Ramakrishna and J. Harrigan, "Defect learning requirements," *Solid State Technology*, January 1989.
27. R. Coleman and P. Chitturi, "Defect reduction strategies for submicron manufacturing tools and methodologies," *SPIE*, Vol. 1392, *Advanced Techniques for Integrated Circuit Processing*, 1990.
28. C. King et al., "Electrical defect monitoring for process control," *SPIE*, Vol. 1087, *IC Metrology, Inspection and Process Control III*, 1989.

29. G. Riga, "Failure analysis, feedback to integrated circuits design and fabrication," *ATFA 77 IEEE,* September 1977.
30. J. Soden and C. Hawkins, "Electrical properties and detection methods for CMOS IC defects," *IEEE Proc. First European Test Conf.,* April 1989.
31. I. Henderson, "A production fab defect reduction program," *Intl. Semiconductor Manufacturing Sci. Symp.,* 1989.
32. D. Liljegren, "Defect reduction strategies for process control and yield improvement," *SPIE,* Vol. 1392, *Advanced Techniques for Integrated Circuit Processing,* 1990.
33. A. McCarthy et al., "A novel technique for detecting lithographic defects," *IEEE Transactions on Semiconductor Manufacturing,* Vol. 1, No. 1, February 1988.
34. D. Bakker, "Applied use of advanced inspection systems to measure, reduce, and control defect densities," *SPIE,* Vol. 1261, *Integrated Circuit Metrology, Inspection, and Process Control IV,* 1990.
35. H. Caludius, "Practical defect reduction in an MOS IC line," *Microcontamination,* April 1991.
36. S. Cooper and C. Geisinger, "Inspection improves production yield," *Electronic Packaging and Production,* March 1988.
37. E. Castel and A. Ray, "An integrated approach to defect detection, analysis and reduction in photolithography," *Microelectronic Engineering,* Vol. 6, 1987.
38. R. Glang, "Defect size distributions in VLSI chips," *IEEE Trans. on Semiconductor Manufacturing,* November 1991.
39. D. Friedman and S. Albin, "Clustered defects in IC fabrication: Impact on process control charts," *IEEE Trans. on Semiconductor Manufacturing,* February 1991.

Chapter

# 4

# Reliability

## 4.1 Introduction

The final chapter of this book is about reliability of integrated circuits (ICs). In the previous chapters, we examined the key modules that are needed to fabricate an IC with multiple levels of metallization. The focus of these chapters has been on multilevel interconnect design, processing, and manufacturing. The reliability program starts early—at the development stage—and continues until the process and products are certified for shipment to customers. In case of a process excursion or change, an evaluation of the reliability impact is done to make sure the customer is continuing to get products of equal or better reliability.

Integrated circuit reliability is affected by its design, fabrication, materials and testing methodologies. In multilevel metal circuit design, interconnect layout and routing are some of the key issues. Choice of materials and their interactions with one another form the crux of the process-related reliability issues. Defects contribute to the degradation of yield and reliability in high-volume IC manufacturing. Finally, IC testing in a manner that is both effective in screening defective parts and cost-efficient is one of the exciting challenges that needs to be tackled.

### 4.1.1 Reliability link

It is important to understand and realize that ICs are one of the basic building blocks of any system that relies on electronics. The reliability of the electronics system is only as good as the reliability of its semiconductor components. Hence, let us examine the impact of component reliability on system reliability. The customer, for example, a computer manufacturer, uses ICs in the manufacture of computer systems. These

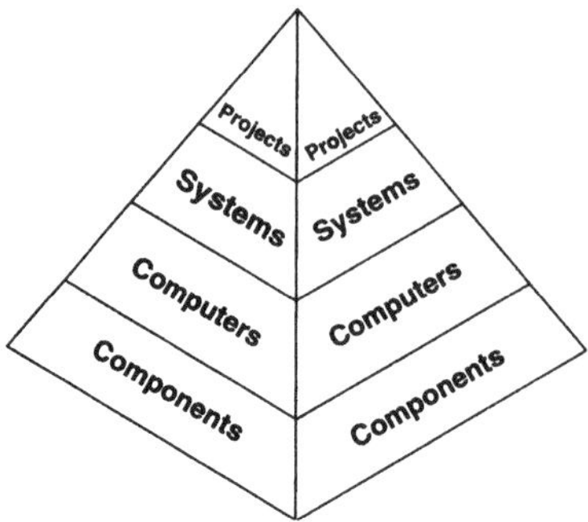


**Figure 4.1** Relationship between IC vendors and customers.

systems are in turn used for various applications in space, education, medicine, etc. It is therefore essential to manufacture products that meet a high level of reliability warranted by their use. The relationship between the component manufacturer and end user is illustrated in Figure 4.1. This figure shows a pyramid of customers and suppliers and examines how each is linked to the success or failure of the other. This pyramid starts with the component manufacturer who supplies components to the computer manufacturer, who in turn supplies computers to the systems developer. All the systems are put together in a major project, for example, the space program that relies on the reliable and effective performance of various systems and subsystems. Of course, there are users at each level of the pyramid.

### 4.1.2 Cost of Q&R

A concern often expressed is the cost associated with ensuring that products meet a high quality and reliability (Q&R) standard. Q&R and cost are not mutually exclusive. Achieving a high level of Q&R often results in higher acceptance rate for products, provided the price of products is right and there is a market for them. This results in higher market share, brand recognition, and loyalty. If products are poor in quality and unreliable, the cost to the customer and the manufacturer associated with this situation will be very high. Customers may return products that they have bought and consequently additional orders for products may stop. The only solution is for the customers to be satisfied with the products they buy and place more orders for the same product because of its superior Q&R. There is one aspect of the Q&R program where the costs can rise significantly: increased reliability testing requirements. The costs involved are due to increased inventory of parts waiting to be tested and also due to increased number of testers and testing time. The solution to this problem lies in ensuring that Q&R is built into the fabrication process starting from product design. Herein lies the reliability challenge.

## 4.2 Reliability Concepts

The design and process challenge is to develop new generations of technology that will enhance the performance of the ICs for an advanced set of applications. The manufacturing challenge is to mass produce devices, cost effectively, using state-of-the-art technology with high line and die yields. The reliability challenge is to keep pace with aggressive technology scaling and simultaneously increase the reliability limits to accommodate constantly more stringent reliability requirements. It is essential to understand these requirements as each comes along with a whole new gamut of issues and concerns. Let us look at the reliability scenario and discuss each of the issues separately.[1–5] Figure 4.2 shows the factors that influence and tend to extend the reliability limits. These will define a new set of requirements necessary to meet the technology challenge.

### 4.2.1 Failure rate

What is failure rate? The basis for reliability lies in mathematics, specifically in the probability and statistics of failures in a sample under investigation. The understanding of reliability mathematics and physics is important since every device has an operating lifetime after which it fails. The reliability concepts involve the determination of the probability that a device will fail at some time in the future from the moment it starts operating. In order to model the failure rate of the devices over time, several distribution functions are used that best describe the reliability condition. The models are based on essentially three basic functions: cumulative distribution function (cdf), reliability function, and probability distribution function (pdf). For each failure mechanism the mean time to failure (MTTF) may be calculated by defining the cdf or the pdf specific to that failure mechanism. Table 4.1 presents these basic functions and the MTTF rela-

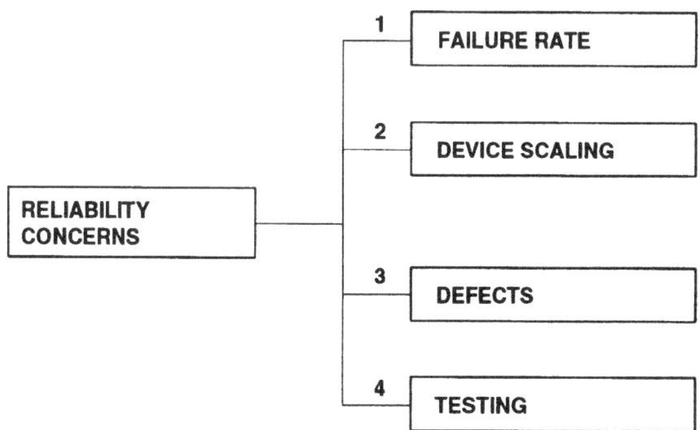


**Figure 4.2** Reliability concerns.

**TABLE 4.1 Basic Reliability Functions[1,2]**

| Function | Equation/properties | Definition |
|---|---|---|
| Cumulative distribution function (cdf) | $F(t) = 0 \quad t < 0$<br>$0 \leq F(t) \leq F(t') \quad 0 \leq t \leq t'$<br>$F(t) \to 1 \quad t \to \infty$ | Probability of failure occurring at or before time $t$ |
| Reliability function | $R(t) = 1 - F(t)$ | Probability of the device to survive to time $t$ |
| Probability distribution function (pdf) | $F(t) = \int_0^t f(x)\,dx$ | Relationship between cdf and pdf |
| Failure rate (instantaneous) | $\lambda(t) = -\frac{d}{dt} \ln R(t)$ | Devices that failed between $t$ and $t + \Delta$ |
| Mean time to failure (MTTF) | $\text{MTTF} = \int_0^\infty tf(t)\,dt$ | Reliability measure—elapsed time before failure occurs |

tionship. Derivations of these equations and a more detailed mathematical explanation are available in the literature.[1,2] A commonly used unit of measure for reliability is a failure unit (FIT). A FIT is defined as one failure/$10^9$ device-operating-hour.

**Intrinsic/extrinsic failures.** In Chapter 3 we discussed how yield is modulated by random and systematic defects. Similarly, reliability failures may be caused by two major components: intrinsic and extrinsic factors. *Intrinsic reliability* may be defined as that component of the overall reliability that is determined by the nature of the design and process used. For example, the use of certain interconnect materials determines the intrinsic reliability window for that device. *Extrinsic reliability* may be defined as that component of the overall reliability that is modulated by defect density, processing conditions, and design rule violations. The goal is to address both these types of reliability concerns. The former is addressed by introducing more robust technology and the latter by implementing a better set of manufacturing practices.

**Components.** What are the components of failure rate (FR)? There are three components to the reliability lifetime of a product: infant mortality, useful life, and wearout. Figure 4.3 shows the distribution of failures over time for any given device and is popularly referred to as a *bathtub curve.* At the start of the reliability testing, a certain population of the sample will fail due to infant mortality. Over time the main failure mode is due to random failure mechanisms, and finally, as the product nears its operational lifetime limit, the failure mode is due to wearout. In each of the three stages, different failure mechanisms predominate and contribute to the degradation of the

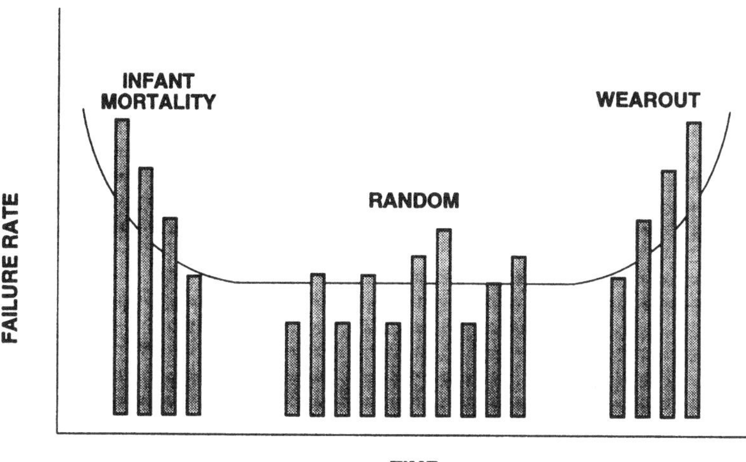


**Figure 4.3** Typical bathtub distribution of failures.

**TABLE 4.2 Bathtub Curve Distributions**

| Distribution | Probability distribution function (pdf) | Applications |
|---|---|---|
| Weibull | $f(t) = \lambda t^{-\alpha} e^{\frac{-\lambda t^{1-\alpha}}{1-\alpha}}$ | Phase 1: infant mortality or early fails |
| Exponential | $f(t) = \lambda e^{-\lambda t}$ | Phase 2: steady state or useful life |
| Lognormal | $f(t) = \frac{1}{\sigma t\sqrt{2\pi}} e^{\left[-0.5\left(\frac{\ln(t) - \mu}{\sigma}\right)\right]}$ | Phase 3: wearout |

product. Building on the reliability probability and statistics theories, the following distributions can be used to best fit the models that can simulate the FRs in each of the three stages of Figure 4.3. A summary of the distribution functions[1] for each stage is given in Table 4.2.

**Infant mortality.** The infant mortality phase is the early failure period. The failures in this period reflect the extrinsic reliability of the devices. They are caused by manufacturing and assembly defects. This is the first stage of reliability screening for defective devices that might have somehow passed the functionality screens. To identify and eliminate the devices with early failures, they are subjected to burn-in testing. *Burn-in testing* involves stressing the whole population (i.e., a batch of devices under reliability test) to electrical stress at high temperature in order to accelerate the temperature-dependent failure mechanisms in a relatively short period of time. Early failures identified by burn-in tests provide valuable information on the extent of manufacturing and assembly defects that are not latent to the process. These failures usually occur in a relatively small percentage of the population. This information can then be used to drive the identification and elimination of the failure mechanisms. For example, processing problems that can cause early failures in ICs are particles and/or contamination, metal defects, dielectric defects, etc. As shown in Table 4.2, the failure rate determined by infant mortality stage may be estimated by using a Weibull distribution function. This may only be used on those parts that fail during the burn-in testing and not on parts that arrive dead at the time of testing.

**Steady state/useful life.** This is a period when there is a fairly constant and low failure rate. This period is the useful life of a device, because after this wearout mechanisms degrade the reliability of the device. Defects that cause failures during this steady period are random in nature and may be unrelated to each other. The overall FR is determined by the intrinsic reliability of the devices under test. For example, failures due to latch-up, gate oxide defects, etc. are generally the failure mechanisms that predominate during this phase. To estimate the failure rate during this phase, exponential distribution functions may be used.

**Wearout.** One of the objectives of every new technology is to extend the useful life of product and push out the *wearout phase*. The wearout phase occurs when a device has gone beyond its normal operating life. Several major failure mechanisms contribute to the wearout of ICs. Some of these are electromigration (EM), hot carrier degradation, and oxide wearout. The focus of much research has been understanding these wearout failure mechanisms and improving the design and the process to make the products less susceptible to wearout. The devices are stressed under different conditions in order to accelerate the failures and to push the parts into the wearout regime. By doing this, one can estimate the failure rate during this period without having to wait for an extended period of time. Lognormal distribution functions are generally used for modeling failures under accelerated testing to simulate wearout. For example, let us consider the influence of temperature on the failure rate. The FR may be expressed as

$$F_t = Ae^{(Ea/kT)} \tag{4.1}$$

where $F_t$ = failure rate (MTTF = mean time to failure)
$A$ = constant
$Ea$ = activation energy

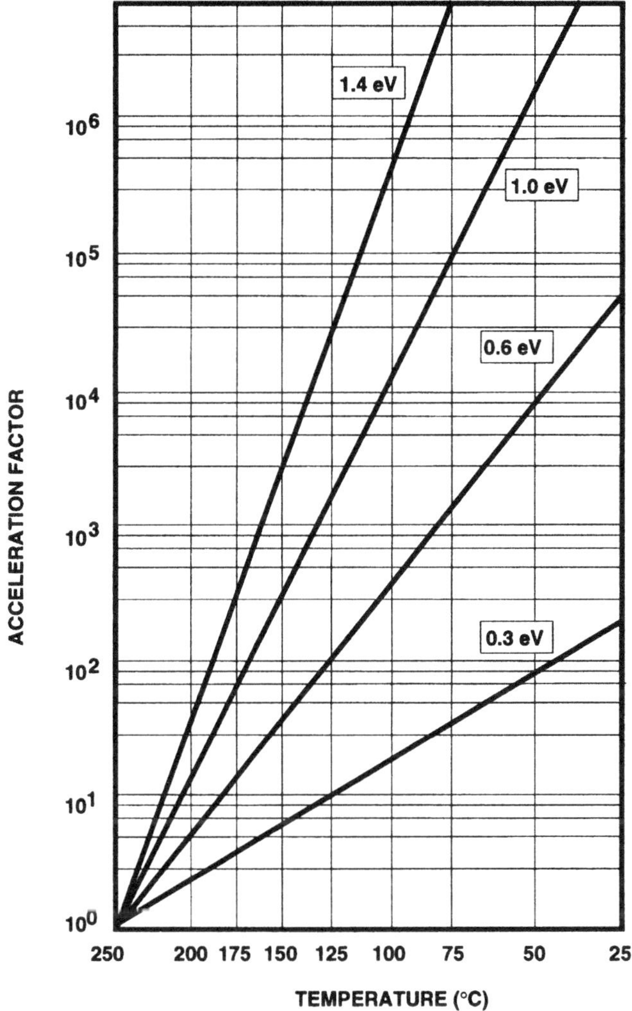


**Figure 4.4** Arrhenius plot illustrating failure rate dependence on temperature.[3]

$k$ = Boltzmann's constant
$T$ = temperature

Using an Arrhenius plot,[3] as shown in Figure 4.4, the FRs over a temperature range may be plotted for different activation energies. So if the maximum operating temperature of a particular device is less than 50°C, then the MTTF for that part may be very long (the assumption here is the failure is not due to infant mortality). Hence it is not practical to spend much time to determine the failure rate of

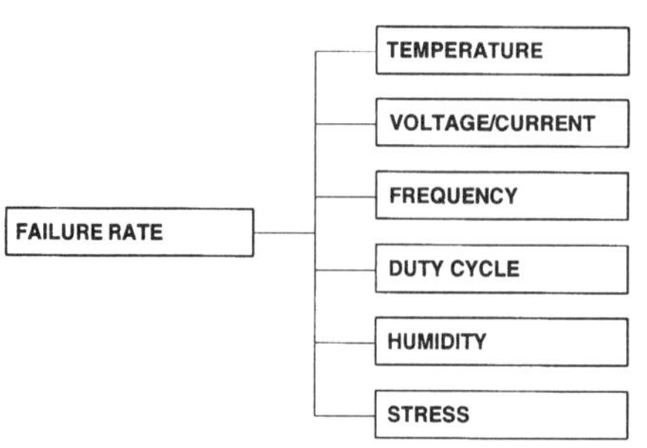


**Figure 4.5** Factors that accelerate failure rate.

devices at low temperatures. Therefore, in order to shorten the testing time, we can measure the failure rate at an elevated temperature and activation energy that best corresponds to the failure rate goal at a lower temperature for the device. Accelerated testing at high temperatures has the assumption that the same mechanisms are controlling failure at the temperature of testing and use. This may put limits on the test temperature. Similar relationships may be obtained for the other factors that influence the failure rate. The FR of a product may be accelerated by a number of parameters. Some of these are shown in Figure 4.5.

### 4.2.2 Customer requirements

Integrated circuit users are demanding lower and lower failure rates from IC manufacturers as devices increase in complexity and the techniques to manufacture them are improved. Mass production of ICs has enabled the cost to decrease as well, thus making the purchase of ICs more cost-effective in the manufacture of computers and other electronic systems.

The failure rates that were acceptable a decade ago are no longer satisfactory and will change again in the next decade. The market is pushing for a greater understanding of the reliability physics associated with the current and future technologies. Market forces are driving IC manufacturers to reduce the failure rate of the chips they sell. It is estimated[6] that by the turn of the century the failure rate requirements could be less than 10 FIT, almost an order of magnitude reduction from the last decade. Such a failure rate may have to be achieved on devices with submicron metal pitches. Stiff competition and demand for the latest generation of products is resulting in tougher reliability requirements on both IC manufacturers and their products.

### 4.2.3 Device scaling

The impact of scaling is felt on all aspects of IC design and fabrication. Consequently, some of the reliability problems are enhanced as

**TABLE 4.3 Scaling Impact on Reliability**

| Reliability problem | Scaling issues that impact failure rate |
|---|---|
| Hot carrier effects | Supply voltage unchanged<br>Features sizes scaled down |
| Oxide breakdown | Supply voltage unchanged<br>Gate oxide thickness reduced |
| Junction breakdown | Supply voltage unchanged<br>Doping concentrations increase<br>Junction depths decreased along with other features |
| Electromigration | Linewidths, contact areas, and step coverage all decrease |
| Electrostatic discharge | Overall decrease in features sizes and oxide/junction parameters results in lower breakdown voltages |

a result of downward scaling of transistor and interconnect dimensions. In some cases the problems are caused by the mismatch of scaling effects. For example, if the power supply voltage remains unchanged and the other parameters are scaled down, the combination gives rise to several problems. Some of the common reliability problems that may be caused by scaling are given in Table 4.3. As the devices are scaled down and new technologies are used, the intrinsic reliability window is constantly tested and reduced, provided the associated reliability problems are solved. The solution may be in improved designs or processes. The intrinsic reliability window may be widened partially by switching to newer materials, for example, the use of aluminum-silicon-copper (Al-Si-Cu) instead of Al-Si for interconnects. The addition of Cu to aluminum has significantly increased the electromigration resistance of the metal interconnects. In addition, the use of tungsten (W) plugs and barrier metals have also enhanced reliability of devices.

### 4.2.4 Defects

Defects in the manufacturing of the devices degrade both their yield and reliability. Defect reduction is a continuous process and not a one-time effort. The defect reduction process is not restricted to one set of processes, but to the whole gamut of IC design, process development, and manufacturing phases. Key to this effort is the application of six-sigma continuous improvement approach.

**Six-sigma approach.**[7] The reduction of feature size and the increase in die size have made the devices more susceptible to higher defect densities, as explained in Chapter 3. Reducing defects through effective manufacturing practices and reduction in process variation is the only avenue left to ensure high yields and reliability. Once the systematic defects have been eliminated, the yield and reliability levels are dependent on the random defect component. In the previous chapters we discussed defect reduction strategies. At this juncture, it is important to revisit the defect reduction strategy and examine how the application of the six-sigma approach is beneficial in bringing about a total quality and reliability improvement. We cannot simply use expensive reliability screens to weed out defective parts; we need to put in place continuous improvement methodologies upstream in the design, process, and manufacturing of ICs. The concept extends to anyone who produces a product. Recognizing and achieving this constitutes total quality control. This is the most cost-efficient way to ship quality parts to the customer.

What is the six-sigma approach? The basis of the six-sigma approach is the normal distribution, as shown in Figure 4.6. Most manufacturing processes have such a distribution. Our objective is to ensure that output parameters are centered and close to the target or mean ($\mu$) value with a low standard deviation or sigma ($\sigma$). If the distribution is normal, 99.73 percent of all the values will be between $\mu - 3\sigma$ and $\mu + 3\sigma$. When the process variation is greater than the specification limits, then there is a definite probability that defects will occur. In the manufacturing process, shifts of the mean by as much as 1.5 standard deviations are possible. In order to ensure the process is robust enough to withstand such shifts in the mean, the six-sigma approach is needed. This approach calls for using $\mu - 6\sigma$ and $\mu + 6\sigma$ as limits for controlling the process. These limits encom-

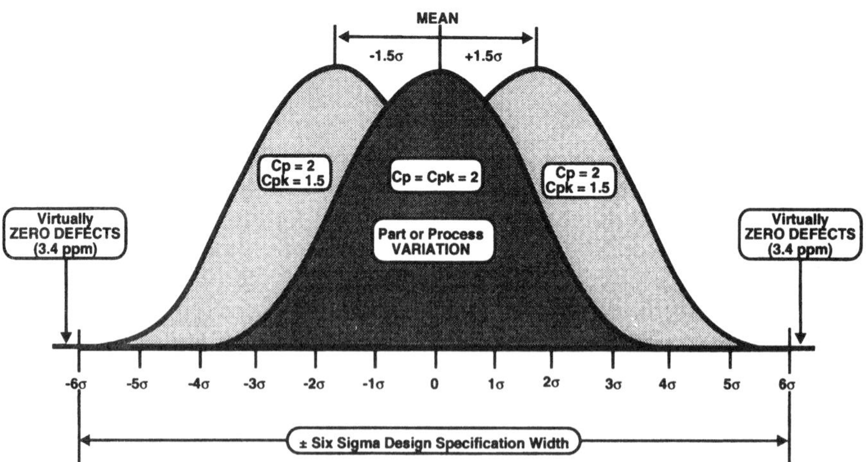


**Figure 4.6** Normal distributions highlighting six-sigma locations.[7] (© *1990 IEEE.*)

pass 99.99966 percent of all values within the normal distribution. If the specification limits are also set with a range of 12 sigma, then the capability index ($C_p$) will have a value of 2. The capability of a process is given by

$$C_p = \frac{U_{sl} - L_{sl}}{6\sigma} \tag{4.2}$$

where $U_{sl}$ and $L_{sl}$ are the upper and lower specification limits, respectively. Another figure of merit that is commonly used to gauge the variation of a process is the $C_{pk}$ value of a process. This measures the deviation of the mean from specification limit as is defined as follows:

$$C_{pk} = \min\,(C_{pu}, C_{pl}) \tag{4.3}$$

where

$$C_{pu} = \frac{U_{sl} - m}{3\sigma} \tag{4.4}$$

$$C_{pl} = \frac{m - L_{sl}}{3\sigma} \tag{4.5}$$

The six-sigma approach suggests that a $C_p = 1$ is not adequate. The process must have a $C_p = 2$ or more. This means that the specification width is twice as large as the process variation ($-3\sigma$ to $+3\sigma$). The obvious advantage in implementing the six-sigma approach is to ensure that defect-free products are shipped to the customer. Let the specification width be represented by $S_w$. Table 4.4 puts into perspective the three different relationships between the specification width, $S_w$, and the process variation ($-3\sigma$ to $+3\sigma$). This table also presents the impact of the three different cases on the quality and reliability of the products. In the second case, one way of obtaining a $C_p > 2$ may be achieved by widening the specification limits. In some cases, this may be appropriate. In most cases, this does not amount to reducing the variation in the process.

### 4.2.5 Reliability and yield models

In Chapter 3 we examined how defects may be generated and how they degrade the yield of the devices. In some cases, these defects may not affect the yield, but may cause a reliability failure during the operational span of the device. As described in the bathtub curve for reliability failures, the extrinsic or early failures could be induced by defects during the manufacturing process. These defects are generated randomly as a consequence of the manufacturing process and remain undetected at the various visual and electrical screening locations.

**TABLE 4.4 Process Variation and Impact on Product**

| Case | $C_p$ | Impact on product quality and reliability |
|---|---|---|
| $S_w = 6\sigma$ | $= 1$ | Fair<br>Some percentage of defective parts produced and affected by shifts in mean value |
| $S_w > 6\sigma$ | $> 1$ | Excellent<br>If $C_p > 2$ then six-sigma methodology is in effect |
| $S_w < 6\sigma$ | $< 1$ | Poor<br>Specification limits are inconsistent with the natural variation of the process |

We can develop a relationship between the random yield models and reliability provided we assume that both yield and reliability loss are due to random failures and are not limited by systematic failures. Let the reliability failure rate or yield be $R_y$ and the die yield due to random defects be $Y_r$. Then

$$R_y \infty Y_r \tag{4.6}$$

This implies that the reliability is dependent on the critical die areas (for defect purposes) and the defect density. This relationship[8] is clearly illustrated in Figure 4.7, which examines the reliability and yield defects. Once such a relationship is established over a long period of time, the reliability performance may be predicted based on the die yield and the associated defect density. This model has a few limitations. For example, the model is more accurate for a batch of lots and is not very accurate for predicting lot level performance. As this model is restricted to random defects only, the reliability picture is limited to this. Defects are also generated during assembly operations, which are not considered here. But this yield/reliability model provides a good screening measurement. Screening parts with low yield and preventing them from being sent for reliability testing saves resources. As the yield/reliability models provide an insight into the defects that may cause reliability problems, corrective actions may be initiated. Both of these are proactive approaches to make reliability a part of the manufacturing process.

### 4.2.6 Testing

How do we ensure that the products we are shipping to customers are reliable? This is a very important question that is being debated and discussed in various workshops and conferences. There is no doubt that customers must get the most reliable product, but this should be managed efficiently. With each new generation of product and process

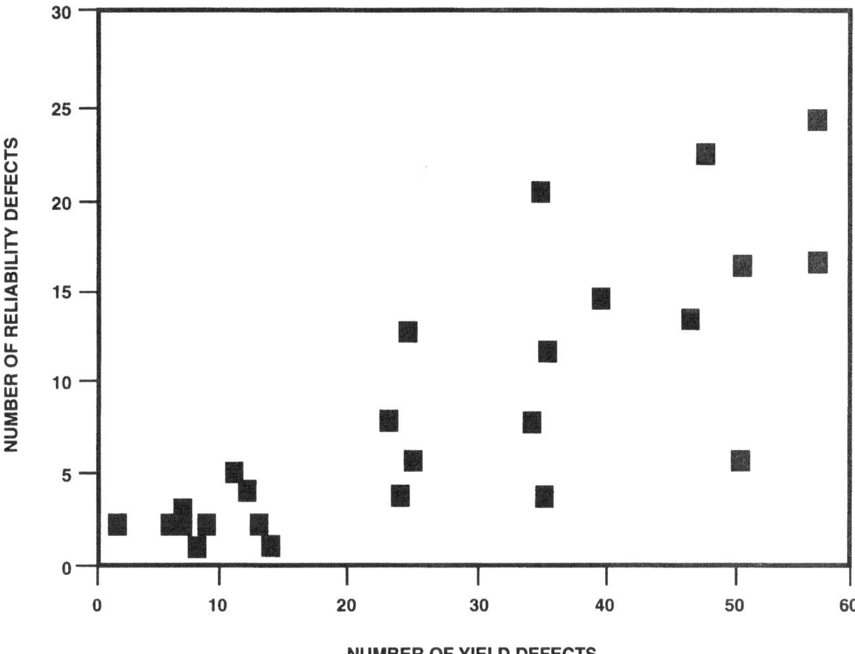


**Figure 4.7** Correlation between yield and reliability defects.[8] (© *1992 IEEE.*)

technology, reliability testing demands are increasing. The management of testing methodology is critical in two ways. First, it is important to ensure that the parts meet the reliability standards and no faulty or marginally reliable parts are shipped to the customer. Second, the cost in conducting these tests needs to be controlled and consistent with the testing requirements. There are three ways that reliability of products may be determined and these are elaborated here.

**End-of-line testing.** Current reliability testing methodology involves testing the products after they have been fabricated and packaged. These tests are screening tests to weed out defective parts and to determine if the failure rates are consistent with the goals. End-of-line or final reliability tests have been designed to evaluate the cumulative effect of the design and all fabrication processes. If there is a failure, then a detailed failure analysis may be necessary to determine the source of the failure. This takes time. If a process excursion was responsible for the reliability failure then the excursion might have passed by the time failure analysis results are obtained. That is a problem of these final reliability tests. They put us in a reactive mode with a major issue in hand. The reactive approach does not provide real-time feedback to the manufacturing process for any correction or modification to the process. However, final tests are suitable to monitor the product population from time to time and during process changes. The results may be compared with previous data. These tests are needed to monitor the wearout mechanisms, such as EM, hot carrier degradation, etc. and will continue to be used for projecting the MTTF and sigma for each failure mechanism.

**Wafer level reliability.** To provide some amount of real time feedback, wafer level reliability (WLR) tests have been considered.[6,9,10] These are not meant to replace end of line tests, but are aimed at providing quick feedback about the process reliability by providing timely information on certain types of process excursion. WLR is accomplished by using the existing monitors in combination with some special reliability structures on the wafers. For example, in the backend (BE) of the process, WLR may be used to monitor the metallization processes and evaluate their impact on EM. This is accomplished by using fast, highly accelerated tests on some special structures. Each of these structures have been designed to address a specific failure mechanism. These tests are still not completely capable of predicting failure rates, but have been shown to flag process perturbations. Similarly, special tests and structures are used for monitoring oxide breakdown, contamination, hot carrier degradation, etc. The WLR techniques are good in the process development phase. In the development mode, special yield and reliability vehicles may be used to develop a process. On these test vehicles, WLR structures may be defined to obtain reliability feedback. The issues are different in a manufacturing setup where the charter is different.

In a high-volume manufacturing environment, there is always a conflict between engineering material, process monitors, and production lots. All three are important and only the last is revenue-earning. There are two issues that face WLR testing. First incorporating this methodology in a manufacturing system. Second, the scribe line real estate is at a premium so that reliability and yield structures in the scribe line are limited. In view of these considerations, the cost of implementing WLR testing has to be carefully evaluated. The concept of WLR—quick and timely feedback on the reliability implications of a process excursion—is excellent. The problems lie in the implementation and obtaining a broad consensus for carrying out.

**Continuous improvement.** There is general consensus in favor of making continuous improvement in all areas of processing and manufacturing. This is attested by the fact that most factories have improvement teams that have active participation from manufacturing operators and supervisors. The mandates are coming from the management. Is continuous improvement the answer to improving reliability? The answer is obviously yes! Engineering effort will always be directed toward addressing process excursions, and manufacturing effort (through improvement teams) will be on tightening the normal variation as much as possible.

The future of reliability testing may be a combination of all three methodologies. End-of-line tests can not be totally replaced with inline tests. Each has a purpose and must be used to effectively monitor the reliability performance of the products. Continuous improvement in all areas is the common thread.

## 4.3 Device Reliability

Device reliability is impacted by several wearout failure mechanisms. The objective in reliability enhancement is to prolong the useful life of the products and fortify the devices from wearout failures. These failures are not easy to detect during wafer processing and become evident out in the field when it is too late to react. Three of the wearout problems that degrade the reliability of the devices are discussed in this section. These are

1. Hot carrier degradation[2–5,11–13,39,50,51]
2. Oxide breakdown[2–5]
3. Electrostatic discharge[4,5,14–20]

The discussion on these various wearout mechanisms will put the reliability issues in perspective as some of these wearout mechanisms are influenced by BE processing. It is therefore important to understand the interaction between transistor degradation and interconnect definition.

Before discussing each of these specific failures, it is appropriate to take a quick look at the location and source of the failures, as shown in Table 4.5. The location is either the gate area in the transistor or the interconnects. The source is either front end (FE) or BE processing. In some cases, both of these may have an effect on the failures. Even though the source of the failure may be FE processing, other subsequent processing may enhance the degradation. This will

**TABLE 4.5 Device Wearout Failures**

| Failure | Location of failure | Source of failure |
|---|---|---|
| Hot carrier | Gate region | FE/BE |
| Oxide breakdown | Gate region | FE |
| Junction breakdown | Gate region | FE |
| Electrostatic discharge | Gate region | FE/BE/external |

become clearer as the discussion proceeds. Device reliability is a cumulative effect. The history of the device fabrication is a reflection of the quality of fabrication, starting from design to manufacturing.

### 4.3.1 Hot carrier degradation

Hot electron degradation has been known for some time to be a problem affecting ICs. As the device dimensions have gone below the micron marker, devices in the submicron regime are severely affected by this reliability problem. The impact of hot carrier degradation is in the form of transistor threshold voltage instability, gate oxide charging, and latch-up.

What are hot electrons or carriers? The reason these carriers are called "hot" is because the electrons gain a significant amount of kinetic energy and inject themselves into the gate oxide, and as a consequence there is charge trapping. The device dimensions have been continuously scaled down over the years without a commensurate voltage scaling. At present a lot of devices operate at 5 V. A new generation of devices has been designed to operate at 3.3 V. If the minimum feature sizes decrease and the operating voltage were to remain unchanged, the electric field in the gate area would increase. If this increase is larger than the transistor operating voltage then hot carrier degradation takes place.

The energy distribution is larger than the energy of the carriers in a thermal equilibrium within the lattice. As the hot carriers gain sufficient energy, they inject themselves into the gate oxide after overcoming the $Si/SiO_2$ barrier, which is approximately 3.1 eV. Let us examine the factors that affect the hot electron injection current. To aid in understanding this, the factors have been broken into external and internal. External factors are electric field and current:

1. The electric field in the region
2. The drain current: number of electrons actually being injected into this high-field region

Internal factors are effects of the external factors that enable the carriers to become hot:

1. Obstacles for electron movement: collision rate of the electrons. The more collisions there are, the more momentum transfer occurs.
2. Mean free path: distance the electrons have to travel before a collision occurs.
3. Barrier height: the energy barrier the electrons have to cross in order for it to be injected.

As a consequence of electron injection, interface states, trapped electron and holes are generated at the $Si/SiO_2$ interface. Their distribution is strongly modulated by the electric fields that are prevalent in the region. Once the electrons are injected, they cause threshold voltage shifts over time. The degradation in threshold voltage is not detected during transistor functional or parametric testing. This degradation manifests itself after the devices have been in the field for some length of time.

There is a difference between n-channel and p-channel hot carrier effects because of the difference in majority carriers in each of the two types of transistors. In n-channel transistors, hot electron degradation results in threshold voltage ($V_T$) shifts, transconductance ($G_m$), and output current. Shifts in these parameters result in reduction of the circuit speed. In p-channel transistors, the wearout also results in parametric shifts. Key to p-channel hot carrier degradation is the reduction in the channel effective length and also in the threshold voltage. The immediate impact of these is observed in the reduction of punch-through tolerance. The bottom line is that unwanted shifts are undesirable.

How do we measure hot carrier degradation? Evidence of hot carrier degradation is observed in the parametric shifts of some transistor parameters. Hot carrier degradation is usually characterized by defining the device lifetime as the time required to reach a certain level of shift in the transistor parameters. Typically the failure criteria may be a percent change in $V_T$, $G_m$, and $I_{Dsat}$ and is device specific.

In order to observe hot carrier degradation, the devices are exposed to accelerated stressing. The external factors that put the devices under stress are the bias (on the gate and drain) and temperature. In n-channel transistors, the worst-case scenario is when the $I_{sub}$ reaches the peak. In the p-channel transistors, the worst case will be when the $I_{gate}$ reaches its peak value.

What are the causes of hot carrier degradation? To understand this, a cause and effect diagram is used to illustrate the salient parameters that cause hot carrier degradation. Figure 4.8 shows the cause and effect for hot carrier degradation. As described in Chapter 2, process-induced charging is a major reliability problem that may be caused by a host of BE processes. Characterization and evaluation of the BE processes for possible hot carrier degradation may be done using structures with different antenna ratios. The antenna structure investigates the combination of charging to breakdown the gate oxide during the plasma process itself and, possibly, the latent defects induced by ionizing radiation during the process. The causes for hot carrier degradation in this diagram are those that are related to BE processing, such as plasma etching. The traps induced by ionizing

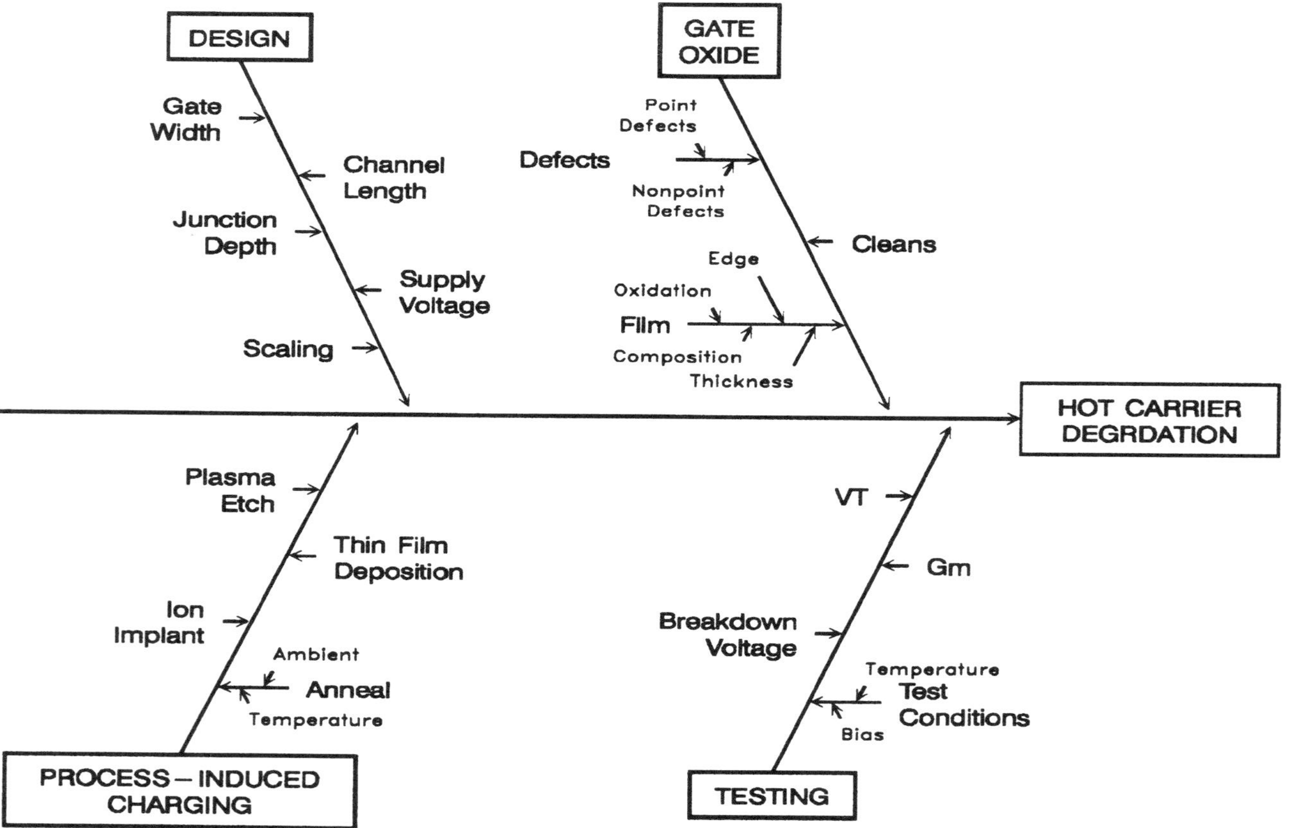


**Figure 4.8** Cause and effect diagram for hot carrier degradation.

radiation may not be fully removed by alloy (anneal) and will contribute to hot carrier sensitivity. This is one of the reasons to move to lower voltage plasma ambients. Ion implant is also a source of process induced charging. The design segment is critical as it impacts several BE processes, which in turn may contribute to the hot carrier degradation. Scaling affects a lot of parameters. Downward scaling of device geometries opens up a new set of variables for the BE processing and therefore may make the devices more susceptible to hot carrier degradation. For example, to improve the gap fill in submicron spaces, new dielectric materials are used. Depending on their moisture content, the device characteristics may vary. Further explanations are provided in Section 4.6. Central to most reliability problems is the quality and integrity of the gate oxide. Defects are a major source of gate oxide degradation. Hence, a separate segment is provided in the cause and effect diagram (Figure 4.8).

How can we prevent hot carrier degradation? As the cause and effect diagram (Figure 4.8) reveals, there are several factors that can cause this failure. The primary thrust in enhancing device reliability is to improve the gate oxide quality and to reduce or eliminate sources of process induced charging. Lightly doped drains (LDDs) are now part of the transistor formation. The LDD reduces the substrate and gate currents and consequently improves the hot carrier lifetime. A lower operating voltage is being used for the next generation of products.

### 4.3.2 Oxide breakdown

Gate oxide quality and integrity are critical elements in ensuring the proper functioning and subsequent reliability of devices in the field. Gate oxide breakdown may be classified into three types of failures depending on their mechanisms. Their degradation is time-dependent.

**Induced defects.** Gate oxide breakdown may occur due to defects in the oxide under normal operating voltages, that is, at low electric fields between 0 and 2 MV/cm. These failures occur in the early stages of the device operational life and have their source in the manufacturing process. The quality of substrates, cleans, and the oxidation process all contribute to the generation and accumulation of defects in the oxide film. The failure rates exhibit a declining trend over time since these defects induce failures that may be screened during reliability testing. In some cases, process excursions may induce gate oxide damage or degradation. In such cases, electrical tests at the end of the process flow may provide adequate screening for poor-quality gate oxides. Defects in the oxide may cause localized weak spots where breakdown occurs. Figure 4.9 shows the causes for defect-induced gate oxide breakdown.

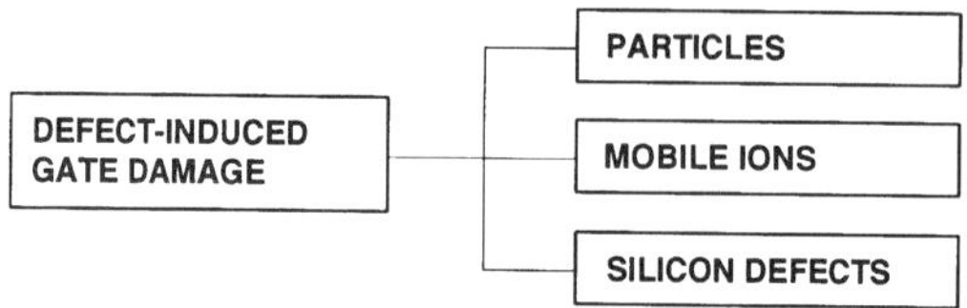


**Figure 4.9** Main defect mechanisms causing gate oxide breakdown.

The defects that induce gate oxide damage or breakdown may be split into three groups. Particles and other contamination that are a consequence of the gate definition process contribute to the degradation of the oxide. In addition to this, mobile ion contamination and control are yet another source of gate oxide integrity problems. Cleans and gate oxidations are carefully monitored to prevent the diffusion of mobile ions. Interlevel dielectrics, such as borophosphosilicate glass (BPSG), help getter $Na^+$ and other mobile ions. The last type comprises defects that are induced in the silicon lattice itself. Stacking faults and dislocations are some of the defects that alter the gate oxide quality.

**Intrinsic breakdown.** The basis for this model is charge accumulation in the gate oxide that overwhelms the oxide after a certain point. Breakdown then occurs through tunneling, which may be predicted by the Fowler-Nordheim (F-N) tunneling emission theory. The typical electric field range is around 10 MV/cm. To characterize this breakdown, we can look at the lifetime of the oxide—the time required for the charge to accumulate in the oxide before breakdown occurs. The critical charge is proportional to the F-N current and also to the time to breakdown. Intrinsic oxide breakdown is primarily modulated by the process technology. For a given process, the intrinsic breakdown should occur well beyond the normal and designed operational life of the device. As shown in Figure 4.10, intrinsic breakdown is modulated by several factors. These factors are based on the process technology and vary from technology to technology.

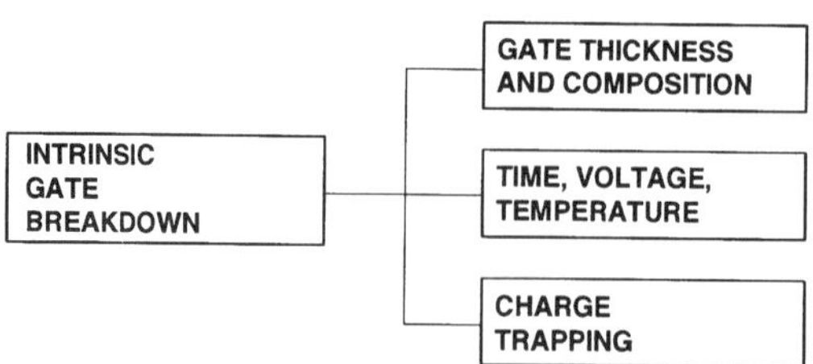


**Figure 4.10** Factors modulating intrinsic oxide breakdown.

**Oxide breakdown.** The reliability concern is the breakdown of the gate oxide due to wearout mechanisms that exhibit time dependent characteristics. This failure exhibits an increasing failure rate with breakdown occurring when the operating voltages approach the intrinsic breakdown levels. As gate oxide thicknesses are reduced, the time-dependent dielectric breakdown (TDDB) becomes a major device reliability limiting issue. Why does the breakdown occur? Breakdown occurs when the electric field across the gate increases and electrons are generated. These inject themselves into the gate oxide, causing the formation of electron-hole pairs. Some of these pairs are trapped in the gate oxide, causing a sufficient charge to build up; the gate oxide breaks down as it cannot hold the excess charge. Electrons may enter the gate oxide by means of quantum-mechanical tunneling effect. The oxide wearout failure mechanism is dependent on a lot of factors and varies from technology to technology. Wearout FRs of the gate oxides may be predicted by accelerating the failures. Figure 4.11 shows the FR due to electric field–accelerated stressing for 100Å gate oxides.[3]

### 4.3.3 Electrostatic discharge

Electrostatic discharge (ESD) and electric overstress (EOS) are major problems that affect integrated circuit reliability. The ESD problem has become worse with the application of advanced technology in the fabrication of complex ICs. The causes for ESD damage are many; the

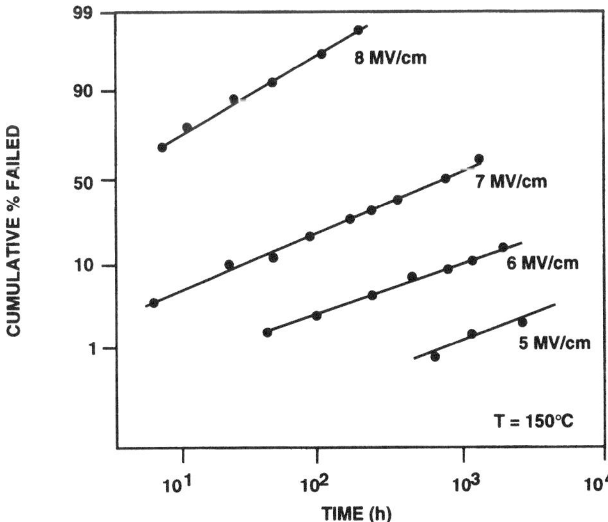


**Figure 4.11** Gate oxide wearout failures due to electric field stresses.[3]

effects are yield and reliability degradation. The solution to the ESD problem lies to a large extent in the recognition that ESD can jeopardize the devices and in the discipline required to reduce it. The cost associated with ESD damage is very high and the economics of this failure mechanism will drive the development of ESD control and prevention. Scaling down of device dimensions makes devices more susceptible to ESD damage. ESD damage may lead to complete device failure by breakdown of the oxides or junctions.

Noncatastrophic failures may also occur that result in threshold voltage or transconductance shifts. Unlike other reliability problems that are time-dependent, ESD failures are primarily an event-dependent failure mechanism. However, by weakening the device, ESD damage may induce a time-dependent component into the failure pattern of the device. Therefore, ESD may be viewed as both an event- and time-dependent failure mechanism. For screening early failures, burn-in tests are used as they are effective in identifying failing devices that are driven by time. ESD failures cannot be screened by burn-in tests as they are primarily event-driven failures.

What are the causes of ESD damage? Let us examine the causes of ESD damage in an IC by using a cause and effect diagram, as shown in Figure 4.12. ESD is the sudden transfer of charge into the device, which is then overwhelmed by it. All the possible areas that can induce such a charge are suspects in an ESD audit of the fabrication or assembly sites. Therefore, ESD elimination programs must start with a careful audit. Periodic audits are essential to check if the factors that cause ESD damage are still present or not.

What are the characteristics of ESD and EOS failures? ESD and EOS can cause three main failure mechanisms in ICs. These mechanisms are illustrated in Figure 4.13. ESD and EOS damage is not limited to the three failure mechanisms mentioned in Figure 4.13. Failures result from ESD when the gate oxide breaks down due to input voltage exceeding breakdown voltage. Breakdown of the oxide is enhanced if the quality of the oxide is not high. Defects in oxide provide a preferential path for leakage currents that can overwhelm the oxide and induce breakdown. ESD damage can also occur in defect-free oxides, especially at the gate edges and corners under the gate metal.[15] Oxide defects are generated from poor processing techniques and controls or equipment malfunction.

Not all gate oxide–related failures due to ESD are oxide breakdown type. Some of them are more subtle and exhibit a time dependence. One such problem is latch-up in CMOS circuits. There are several causes for latch-up in these circuits. CMOS circuits are defined with dual wells (n and p) in order to form both n- and p-channel transistors. Latch-up results when parasitic bipolar transistors are formed

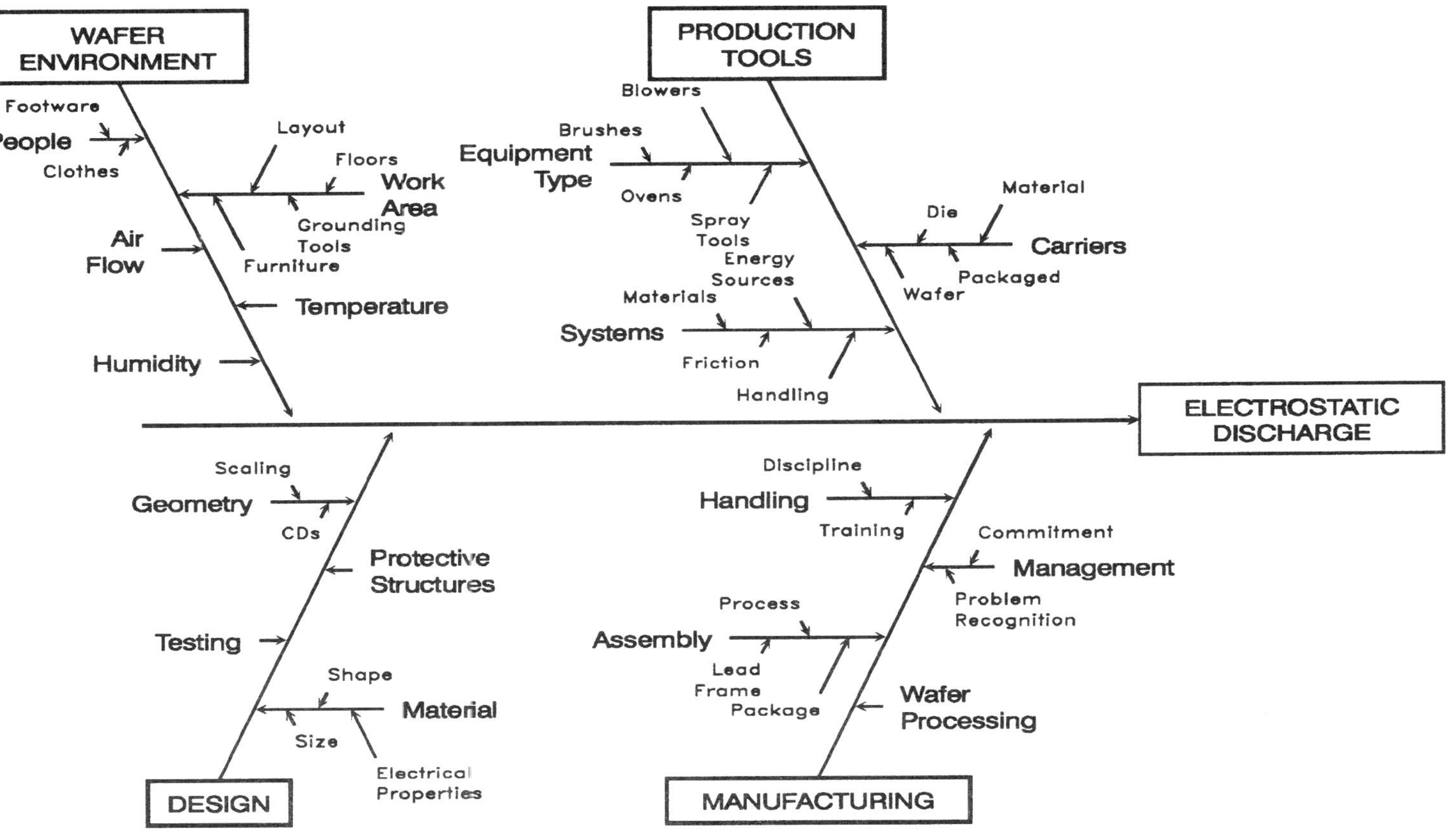


**Figure 4.12** Cause and effect diagram for electrostatic discharge.

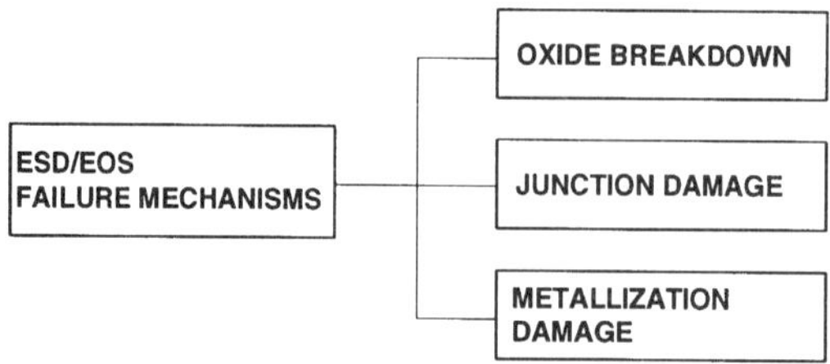


**Figure 4.13** ESD/EOS failure mechanisms.

unintentionally at the edges of the wells. These transistors can cause shorts to occur which therefore latch-up and damage the device.

Junction damage is another major failure mechanism that results from ESD stress. This is observed as a degradation in the I-V characteristics of the device. Junctions experiencing high-power densities are susceptible to junction breakdown as well.

Under a high-stress situation, excessive localized joule heating may melt the interconnects and cause severe damage to them. Depending on the process technology, metallization damage may also cause junction spiking (eutectic formation) and degrade the junctions. In such a case, the Si lattice structure is deformed and junction leakage may be the result. The usual metallization damage as a consequence of concentrated joule heating is the melting and vaporization of the metal interconnects. This obviously results in opens.

What models explain ESD phenomena? There are several models that explain ESD-induced device degradation. Some of these are illustrated in Figure 4.14. The human body model is based on the amount of charge a human body can store. This stored charge can discharge through the device when contact is made with a human body. Standard values that are widely accepted for this model are 100 pF for the capacitance and 1500 ohms resistance. Discharge through the circuit is governed by RC decay. The charge device model may be applied in a more automated environment where the charge transfer takes place during contact movement of the packaged devices against

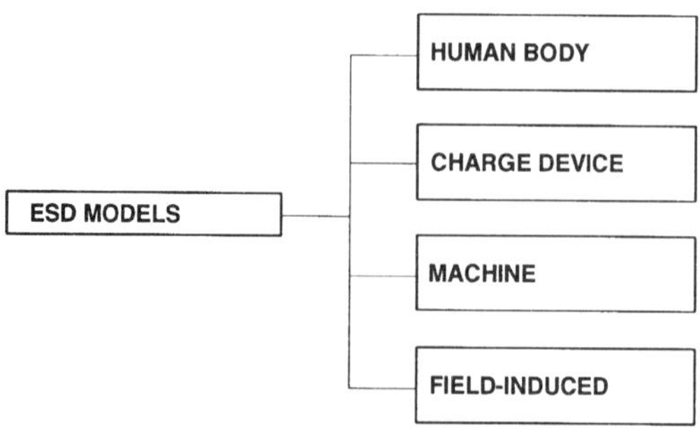


**Figure 4.14** ESD models.

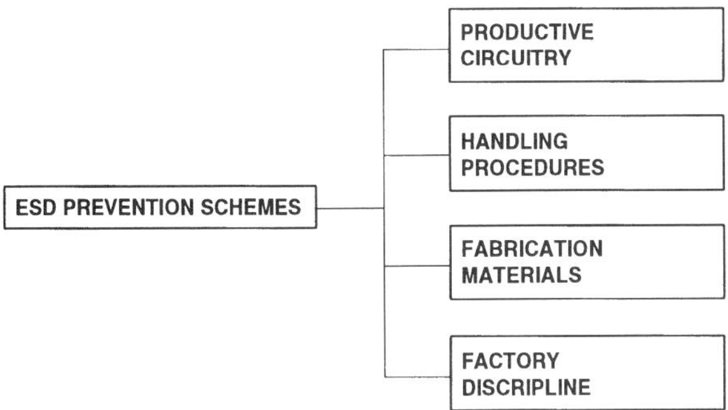


**Figure 4.15** ESD prevention schemes.

other materials. The devices may be destroyed by a rapid discharge of stored charge that has been accumulated on the packaged part. In some cases, an electric field can induce a charge on the device; if it is sufficiently large it may cause ESD damage. Such a situation is explained by the field-induced model. Yet another model that is widely used is the machine model. The static sources may be wafer carriers, carts, or any isolated conductors.

How can ESD damage be prevented? It is often said that ESD prevention must be a way of life in the factory environment. It is not just limited to the IC fabrication or assembly areas. Customers of ICs need to be educated about ESD damage, as accidental or intentional handling of the devices in systems can cause ESD damage. Figure 4.15 highlights some of the salient ESD/EOS preventive schemes. Protective circuitry can be designed to prevent the surge of charge from damaging the gate oxide. This has to be coupled with improved handling procedures in every area of IC fabrication, assembly, and testing. New fabrication materials, such as wafer carriers or package carriers that do not store charge, are some of the areas for development. The last and most important preventive measure is discipline.

## 4.4 Reliability of Dielectrics

Dielectric deposition, planarization, and contact etching are three of the main components of a multilevel interconnect system. In Chapter 2 we discussed the deposition, planarization, and dielectric etching issues in detail. Some of the reliability wearout mechanisms, such as hot carrier degradation and EM, are influenced by the nature of the dielectric material used and its influence on the gate region or on the metal interconnects. In the various cause and effect diagrams that are presented in this and other chapters, dielectrics play a very important role in determining the overall device reliability.

**TABLE 4.6 Dielectric Issues that Impact Reliability**

| Issue | Source | Reliability impact | Remarks |
|---|---|---|---|
| Gap fill | High aspect ratio | Speed | Capacitance/cross talk |
| Stress | Deposition | Metal voids | EM, Stress migration |
| Moisture | Film | $V_T$ shifts | Hot $e^-$ degradation |
| Process-induced charging | Equipment/process | $V_T$ shifts | Hot $e^-$ degradation |
| Delamination | Mismatch between films | Cracks | Pattern deformation |
| Corrosion | P doping (BPSG) | EM, corrosion | P doping used for gettering |
| Step coverage | High aspect ratio | EM | Planarization technique |
| Defects | Equipment | Shorts/opens | Particles |

EM = electromigration.

Table 4.6 puts the various issues associated with dielectric deposition and their planarization in perspective. The issues may be the main or side effects of the dielectric processes as a whole. For example, gap fill may be considered as a main effect (step coverage improvement) of the dielectric deposition as it depends on the film type, deposition conditions, and the aspect ratio. The main effect is something that is a direct result of the process and the side effects are the secondary effects on various yield and reliability parameters as a consequence of the process. The side effect of gap fill occurs when voids are formed and alter interconnect line-to-line capacitance. Another example of this may be found in the application of BPSG as an insulator. In using BPSG, we dope the film with a certain percentage of P to getter mobile ions. The side effect of this may be metal corrosion. The bottom line is that the whole interconnect system must be visualized as one comprehensive unit, and all the causes (direct or indirect) that degrade yield and reliability should be evaluated.

### 4.4.1 Impact on device reliability

As devices are being scaled, several dielectric materials and novel planarization techniques are being adapted to make use of multiple levels of metallization possible. Dielectric deposition and planarization can contribute to device degradation through charge build-up, trapping, and radiation damage. For example, passivation layers, such as silicon nitride, have long been used to protect the finished devices. This nitride film protects the devices from moisture and mobile ion contamination. The latter two degrade the device performance.

The main form of device degradation due to dielectrics has been observed to be hot carrier effects that make its performance unstable and less reliable. Therefore, it is only appropriate to discuss hot carri-

**TABLE 4.7 Hot Carrier Degradation from Dielectrics**

| Causes | Hot e− effects | Remarks |
|---|---|---|
| Moisture | Degrades | $V_T$ instability. |
| Alloy (anneal) | Improves | Removes radiation damage and drives moisture out. Temperature must be compatible with metal interconnects. |
| Radiation damage | Degrades | Process-induced charging and ionizing effects due to radiation, e.g., plasma etching. |
| Film type | TEOS, silane, SOG, silicon nitride | Moisture; generation of more bulk and interface traps. TEOS, SOG, $H_2$-rich plasma nitride cause device instabilities. |

SOG = spin-on-glass; TEOS = tetraethylorthosilicate.

er effects in a device due to dielectric deposition and planarization processes. As a starter, a brief summary of the various factors that may induce hot carrier degradation due to dielectric deposition and planarization is presented in Table 4.7.

Moisture penetration is a major problem that affects device reliability. It has been reported[50] that hot carrier degradation is enhanced by the presence of H and $H_2O$ into the gate oxide. These react with the oxide and form hydride or hydroxyl bonds in the oxide. Such a reaction generates centers for electron trapping. In order to improve the gap fill properties during dielectric deposition, TEOS-based oxides have been used. In addition to these oxides, spin-on-glass (SOG) has been widely used for planarization. This has been found to generate mobile positive charges and therefore degrade reliability. These oxides also absorb moisture that may eventually make its way to the gate region.[40] This causes threshold voltage instability. Moisture penetration has other detrimental effects, such as corrosion and voiding. These will be discussed in the Section 4.5.1, below.

The solution to moisture-induced device instability requires a compromise between selecting a dryer film and its step coverage properties. To improve gap fill properties, films like SOG and TEOS-$O_3$ are widely used, but they cause moisture related device failures. As a compromise to solving the reliability issues, TEOS-$O_3$–based oxide has been limited to filling the gaps; a dryer film, like plasma-enhanced TEOS, is used for providing isolation. In other words, a combination of films may be used that offer good deposition characteristics and low moisture content. A further extension of this concept is the application of a dielectric barrier layer. This barrier layer helps prevent moisture in wetter films from penetrating the gate region.[41] The water trapping barrier layer improves hot carrier tolerance. To investigate this, a comparative study[41] of plasma TEOS oxide and electron cyclotron resonance (ECR) plasma oxide was performed. The data suggests that ECR plasma $SiO_2$ is effective in preventing hot carrier degradation associated with moisture penetration from the interlevel dielectric films. The reason proposed for this is that dangling bonds in the oxide film getter the moisture and prevent it from diffusing.

Development of low moisture content films with good gap fill properties is essential. In a recent study,[42] it has been shown that the deposition condition can be varied to obtain a dryer film. Moisture retention in TEOS oxides can be modulated by refractive index and compressive stress in the film. These may be used as a good inline monitor and also as a characterization tool for developing low-moisture content dielectric films.

Alloying or annealing wafers helps improve hot carrier performance. Annealing wafers is usually done around 400°C due to the low melting point of Al. Annealing has two distinct advantages.

1. Annealing helps drive out moisture from the films. If moisture penetration is a problem for a particular dielectric, a prepassivation anneal is sometimes more appropriate than a postpassivation anneal. The passivation layer may act as a barrier to moisture and therefore prevent the outward diffusion of moisture from underlying films.

2. Annealing in a hydrogen ambient helps in neutralizing trapped charges. The presence of trapped charges may be due to a host of reasons, such as radiation damage that can occur during BE processing. Figure 4.16 shows effectively the effects of ionizing radiation and the ability to remove the damage with postmetallization anneal (PMA).[39] The data was compared with samples that did not receive postmetallization thermal treatment. Wafers were exposed to x-ray radiation. The ability to remove radiation damage may limit our device performance in the deep submicron regime. The only saving factor is that the gate dielectric will be thinner and thus the volume in which traps are induced will be smaller.

Backend processes can induce charging as there are several steps that expose wafers to high levels of radiation. As charging is a cumulative effect, use of more and more metallization layers poses a greater reliability risk. Here again, the compromise lies in developing dielectric deposition, planarization, and etch processes that satisfy design and process specifications, but at the same time do not contribute to device instability. The problems are not just restricted to process parameters, but also to the choice of materials.

To illustrate differences in materials, a recently published study comparing TEOS and silane oxide suggests that TEOS oxide degrades device reliability more than silane oxide.[51] This is explained by the

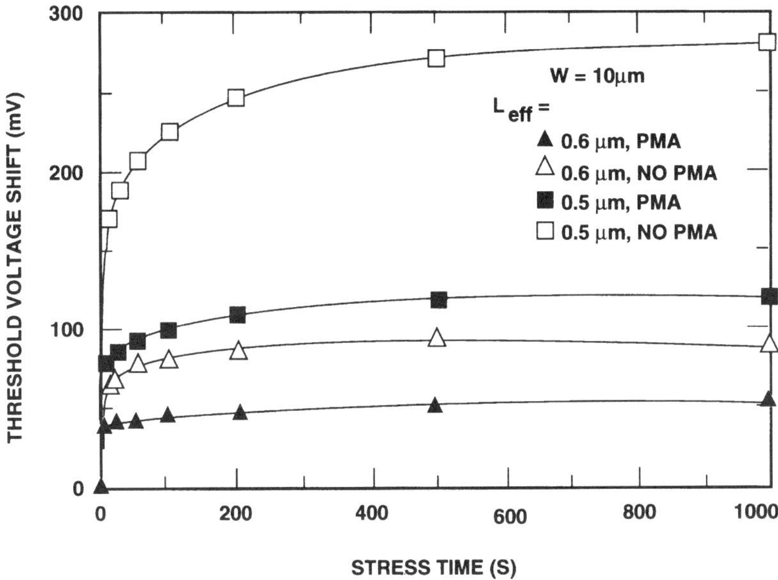


**Figure 4.16** Impact of postmetallization anneal on radiation damage.[39] (© *1992 IEEE.*)

fact that TEOS oxide generates a higher number of bulk and interface traps. It was also found that films that are Si-rich reduce hot carrier degradation.

### 4.4.2 Impact on interconnect reliability

Insulating layers impact the interconnects above and below them. Figure 4.17 shows the salient issues with dielectric interaction with the interconnects. Interconnect failure mechanisms are primarily EM and stress-induced migration.

**Gap fill/capacitance.** As line dimensions shrink, filling narrow and high-aspect ratio gaps with a dielectric film becomes a problem. This problem has two components to it. First, poor gap fill may give rise to poor metal step coverage, which in turn contributes to EM degradation. Second, poor gap fill leads to void formation in the dielectric.

Let us focus on void formation. The void problem has been somewhat addressed to date by using TEOS-based films, with superior gap fill properties, instead of silane. In addition to this, multiple deposition/etch sequences are used to fill the gaps. Voids become a serious issue as metal pitch is reduced. A detailed reliability study is needed to evaluate the impact of voided dielectric on device reliability and speed performance. It is believed the reliability impact of this issue will be in the alteration of capacitance between metal lines. During subsequent pro-

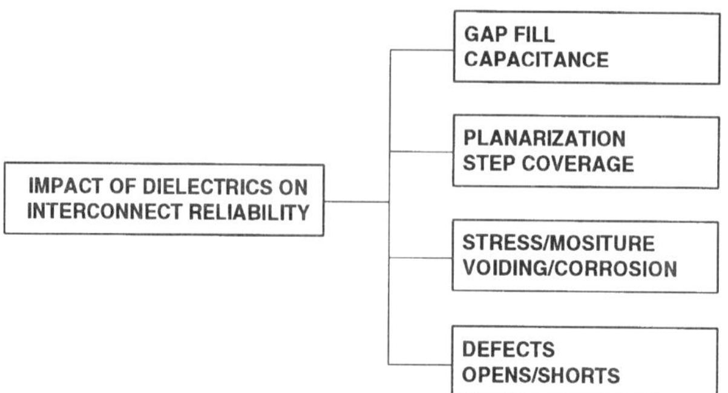


**Figure 4.17** Impact of dielectrics on interconnect reliability.

cessing, voids may be exposed and filled with contamination that could alter the capacitance between metal lines. Unaccounted or unanticipated capacitance between metal lines degrades the switching speed and also causes cross talk between neighboring signal lines. This will result in faulty operation if not fixed. Planarization is an integral part of the dielectric process and the objective is to provide a surface that is smooth for metal interconnects. Step coverage of metal lines over steps and contacts modulates EM performance of the devices. Poor step coverage leads to increased EM fails.

But what about the final passivation layer? Topography of the final metal layer is very severe, primarily due to the thickness of the metal layer (even though it is over a planarized surface). Unplanarized passivation layers could create several problems when the die is packaged, such as thin-film cracking. This affects underlying metal interconnects. It has been shown that planarization of the passivation layer, using a SOG intermediate layer has greatly improved reliability of EPROM devices.[38] Figure 4.18 compares the failure rate on samples with and without SOG passivation planarization.[33] This suggests that passivation planarization significantly helps in lowering the FR.

**Stress/voiding.** Stress-induced voiding is a major problem that affects the Al alloy interconnects. Stress/void formation in passivated metal lines is due to the thermal expansion mismatch between the metal and the silicon substrate. These voids can grow and move around the length and breadth of the metal line. Voids lead to metal line opens and also EM failures. Stresses in the dielectric film may cause thin films adjacent to each other to delaminate, causing interconnect ruptures and ultimately degradation of the functionality and/or reliability of the device.

**Moisture/corrosion.** Earlier in this section we examined how moisture can degrade device reliability. Moisture in films has other detri-

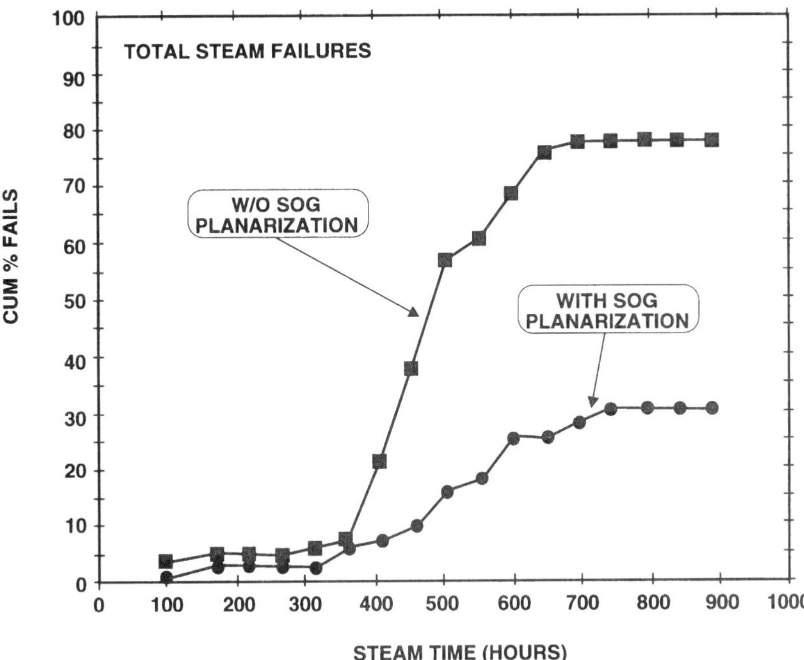


**Figure 4.18** Failure rate reduction by planarization of passivation layer.[33] (© *1989 IEEE.*)

mental effects on interconnect reliability. Firstly, voiding in metal lines can be initiated by moisture in the dielectric film. For example, it has been reported[49] that moisture in SOG films contributes to void formation on Al-Si metal lines. During the cure process the moisture outgasses from the SOG film and produces a pressure cooker effect on metal lines. This causes stress to build up in the metal line resulting in void formation. Secondly, moisture has the potential for starting corrosion in Al.

### 4.4.3 Impact on package reliability

We discussed above how the dielectric layers can alter the reliability performance of the devices by affecting the gate region and the interconnects. Dielectric layers, specifically passivation layers, are also crucial in ensuring that the die can withstand stresses when encapsulated in a package. This aspect of dielectrics is important as a significant amount of fabrication and assembly investment has gone into encapsulation of the device. Failure at this stage is very expensive.

Downward scaling of critical dimensions has required thinner passivation, as stress-induced voiding in Al interconnects is a real problem. Thinning the passivation has led to poor step coverage in narrow spaces. At the corners of these steps, passivation stresses (compressive) are high and can form microcracks at deposition directly or during the passivation etch process. Cracks in passivation are detrimental to device reliability as they offer paths for moisture to penetrate. Once moisture penetrates the device, it is susceptible to corrosion and transistor instability.

Another major problem associated with passivation and its interaction with the package in which the die is enclosed is thin-film cracking.[43–46] As the level of device integration and die size increases, thin-film cracking becomes a major issue. Many factors can combine to induce thin-film cracking. Thin-film cracking results in the deformation of the metal lines, passivation cracking, and even delamination of interlevel dielectric materials. Cracking occurs predominantly when the packaged device is subjected to temperature stress and cycling. The model that best describes this problem is based on the mismatch of the coefficient of thermal expansion between the package material and passivation during thermal stress. During this thermal stress, strong shear forces arise as a result of the differing thermal coefficient of expansion for the two materials. These forces are large enough to distort the metal pattern underneath the passivation layer. Generally, passivation cracking originates at the corners of the die and propagates toward the center. Figure 4.19 shows the cause and effect diagram for thin film cracking. Let us review some of the salient causes from this cause and effect diagram.

Cracking originates at the die corner. Therefore the metal routing layout has to be resistant to the shear forces. Various anchoring structures may be placed at these locations to prevent the propagation of the cracks toward the center of the die. These anchoring structures improve the adhesion of the molding compound to the die and thereby preventing any movement.

Hard passivation films such as silicon nitride offer high resistance to the shear forces exerted by the plastic package during temperature cycling. In addition to this, if the passivation is thin the likelihood of observing thin film cracking is high. To prevent this from happening, the passivation layer must be decoupled from the shear forces. This may be accomplished by planarizing the passivation layer and ensuring that the thickness of planarizing material is sufficient to offer a path for the dissipation of the shear energy. When this path is obstructed by the underlying metal topography it is assumed that cracking takes place. When the path is over a metal pattern that is underneath passivation, the chances of thin-film cracking are minimized. For example, planarization of the passivation may be achieved by using SOG as an intermediate material. Instead of SOG planarization, a thick polyimide may also be deposited over the nitride; this polyimide is soft enough to absorb the thermal shocks caused by plas-

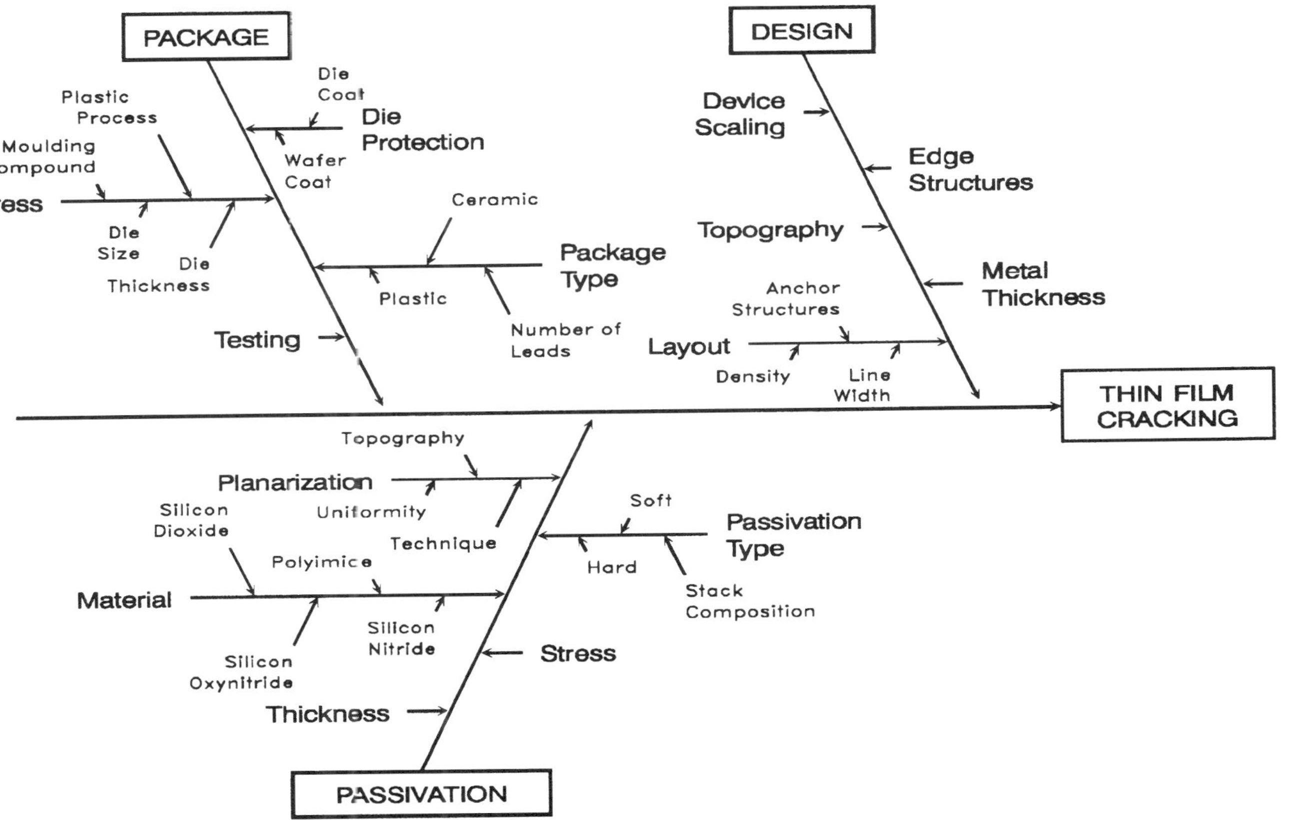


Figure 4.19 Cause and effect diagram for thin-film cracking.

tic packages. In some cases, a silicone gel type of material may also be deposited over the die during assembly steps prior to encapsulation. This serves the same purpose of protecting the die from the plastic package. The best process has to be one that is most cost-effective and capable of withstanding all the reliability tests.

The problem of thin-film cracking occurs in plastic packages as they are subjected to temperature cycling. This builds a significant amount of stress in plastic packages that acts against the passivation. One avenue for package development is in the use of low-stress plastics that are better suited for temperature cycling. The composition of the molding compounds and their treatment are just some of the factors that can be modulated. Thick dice are more susceptible to the cracking problem. Reducing die thickness helps relieve some of the stress that develops in plastic packages.

## 4.5 Reliability of Interconnects

As described in the previous chapters, metal interconnects occupy a large amount of Si real estate. Success in fabricating devices with submicron process technologies requires addressing yield and reliability problems associated with metal interconnects. There are several interconnect reliability problems that affect functionality and lifetime of the devices. Figure 4.20 highlights these problems. Corrosion and hillock formation were discussed in Chapter 3 and will not be discussed here. In this chapter we will examine EM and issues pertaining to metal cracks and voids.[21–49]

### 4.5.1 Electromigration

One of the most dominant failure mechanisms that degrades the reliability of metal interconnects in an IC is EM. A typical EM-induced failure is shown in Figure 4.21. This problem has been the subject of

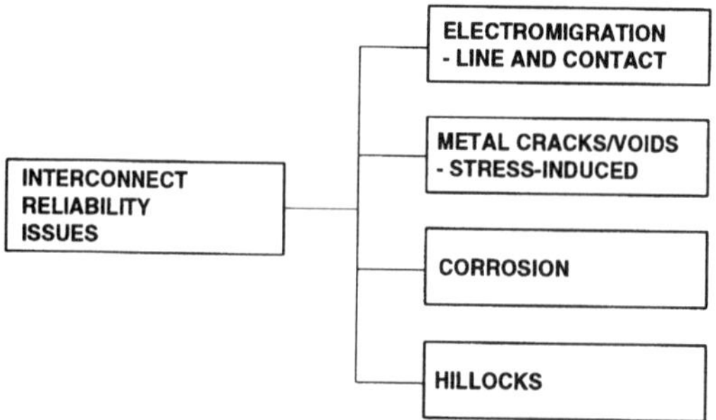


**Figure 4.20** Interconnect reliability issues.

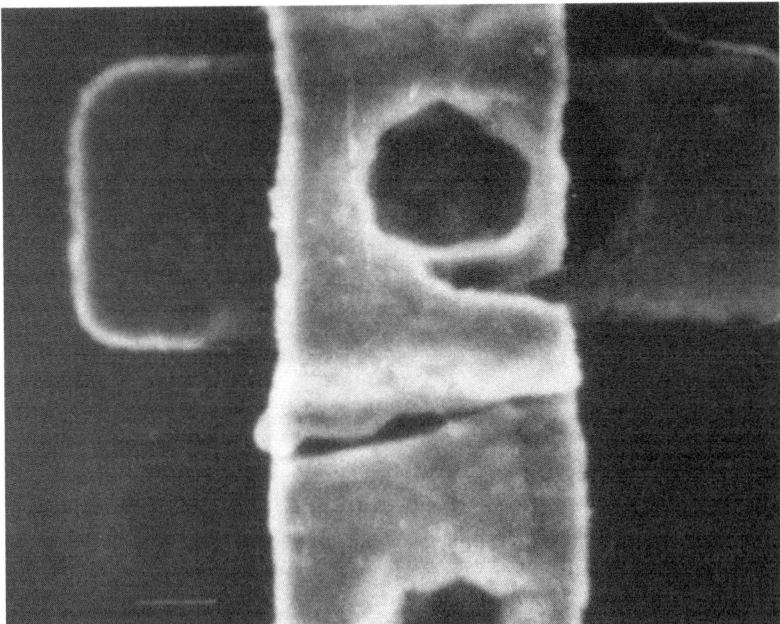

**Figure 4.21** Electromigration-induced failure in Al interconnects.

intense study for quite some time. A thorough understanding of interconnect reliability is needed as we progress into higher levels of integration and use multiple levels of metallization. Ensuring a high level of reliability becomes complicated as interconnect density increases and metal pitch decreases.

What is EM? Electromigration is the transport of ions in the conductive material as a result of current passage through it. As the current (movement of electrons) passes through the conductor, an "electron wind" is generated. This electron wind is responsible for displacing ions of the conductive material from their original location. This displacement occurs because of momentum exchange from the movement of electrons due to the application of the electric field. The displacement causes accumulation and depletion of material. In areas where there is accumulation of material, metal shorts are the main failure mechanism. In areas where there is depletion of material, opens are the main failure mechanism.

Figure 4.22 is a transmission electron microscope (TEM) image[35] of the depletion and accumulation effects of EM in an Al conductor. As the electrons flow across the metal interconnect, the depletion of the conductive material reduces its cross-sectional area. This reduction causes current crowding and an increase in the current density, thus accelerating the material depletion process. The material that has

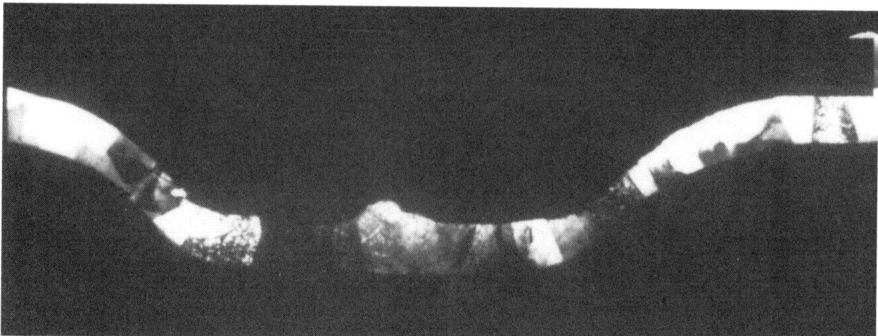

**Figure 4.22** Depletion and accumulation effects of electromigration in an Al conductor.[35] *(Courtesy Materials Research Society.)*

been depleted has to go somewhere and is downstream of the electron flow. The depleted material accumulates in the form of metal hillocks.

In a multilevel interconnect system, all levels of interconnects are susceptible to EM failures, which is an irreversible process. When the failure occurs on the top interconnect layer, the passivation layer on top of it may be fractured by the formation of hillocks thus allowing moisture to penetrate and cause corrosion. The EM failure mechanism may be accelerated by increasing the current density ($J > 10^5$ $A/cm^2$) and also by increasing temperature.

The temperature the metal interconnect experiences during EM stress may vary depending on the function of the interconnect (i.e., metal 1, 2, etc.) and the operating temperature of the circuit (the latter has a direct influence on the temperature of the metal lines). In addition, the temperature of the metal interconnects may be elevated due to poor thermal dissipation and/or an increased self-heating as a result of current crowding. Current crowding may be due to poor step coverage of the metal interconnect or due to high resistivity via metal.

**EM cause and effect analysis.** Several parameters can decrease the EM resistance of the metal interconnects. Figure 4.23 shows a cause and effect diagram that presents the overall picture of the parameters that affect EM. Table 4.8 provides the direction of the impact of the various parameters on EM performance and complements the EM cause and effect diagram. Column 1 shows some of the key parameters that impact the EM performance of the metal interconnect. Column 2 shows the change required to alter EM performance, that is, the MTTF. This change is indicated by a "+", indicating an addition of material to the Al interconnect. The up and down arrows are used to indicate either an increase or decrease in the magnitude of the parameter. An up arrow, "↑", under the EM column means an improvement in the performance and "↓" means a degradation. In some cases where there is no significant impact in the EM perform-

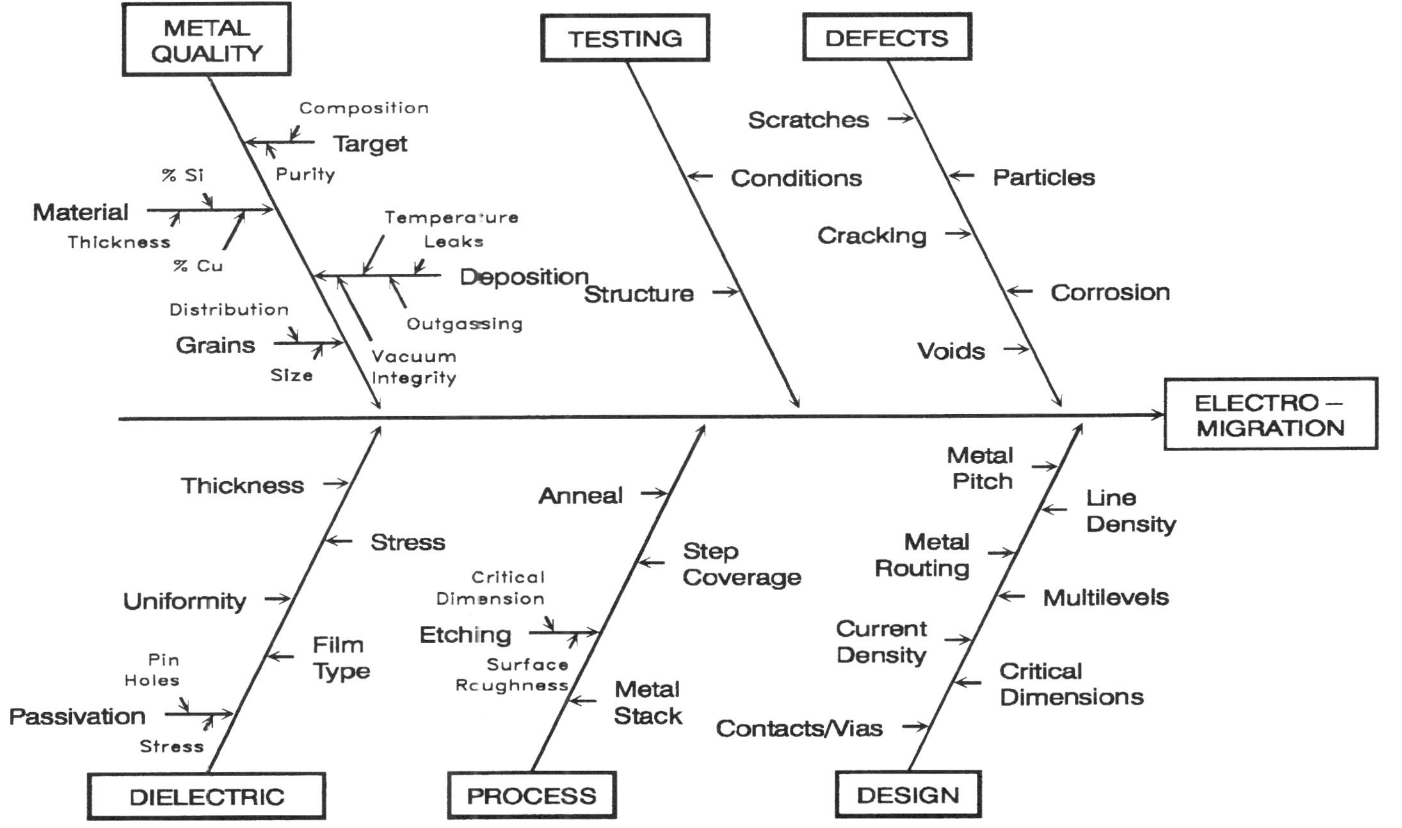


**Figure 4.23** Cause and effect diagram for electromigration.

**TABLE 4.8 Impact on EM Performance**

| Parameter | Change | MTTF | Remarks |
|---|---|---|---|
| Composition | + Cu | ↑ | 0.5–4% Cu is used<br>Improves contact EM |
| Impurities | + $N_2$ | ↓ | $N_2$ traces during metal deposition |
| Grain size | ↑ | ↑ | Larger grain size reduces grain boundary diffusion paths |
| Line width | ↓ | ↓ | Until a critical width is reached |
| Line length | ↑ | ↑ | EM improvement saturates after certain length |
| Thickness | ↑ | — | No impact for constant J |
| Step coverage | ↓ | ↓ | Leads to current crowding; smaller grain size at step |
| Stress | ↑ | ↓ | Stress induces voids |
| Passivation | ↑ | ↑ | Increases stress |
| Barrier metals | + | ↑ | Redundant current path |

ance, a "—" is used to indicate this. Column 3 indicates the most salient remarks. The various parameters are discussed in greater detail below.

**Composition.** The most important contributor to EM performance is the metal quality itself. It is imperative that all processes that ensure a high level of metal quality are effectively managed and controlled. In order to improve the EM resistance of the Al interconnect, Al alloys have been used. The alloys have been formed by adding either Si or Cu and sometimes both. The addition of Si has been primarily for the purpose of improving contact EM. The typical concentration of Si in Al is 1 percent; this is needed to prevent junction spiking during the Al and Si alloy processing. Saturating Al with Si prevents Si consumption from the contact areas, and therefore prevents junction spiking. Junction spiking may be overcome with the use of barrier metal layers. However, with the use of contact plugs, the need for Si-based Al alloy is not necessary. Besides, junction spiking, the addition of Si to the Al, prevents Si precipitates in the contacts. These precipitates decrease contact area and degrade EM performance.

Addition of Cu to Al has greatly enhanced EM performance of Al interconnects. The MTTF has been shown[47] to increase by a factor of 10 over that of pure Al with the addition of Cu. Table 4.9 compares[47] MTTF of various metals for a specific set of test conditions. In generating the data in Table 4.9, EM testing was done at 150°C for the specified current densities. The failure rate was set at 50 percent

**TABLE 4.9 MTTF of Typical Metals and Alloys**[47]

| Type | Current density (A/cm$^2$) | MTTF (h) |
|---|---|---|
| Al | 1.95 E6 | 660 |
| Al-1%Si | 1.30 E6 | 2,100 |
| DC bias Al | 1.00 E5 | 153 |
| Al-1%Si/TiW/Al-1%Si | 1.00 E6 | 1,450 |
| Al-1%Ti-1%Si | 2.49 E6 | 1,500 |
| Al-0.5%Cu | 2.50 E5 | 6,000 |
| Al-0.5%Cu-1%Si | 2.50 E5 | 4,500 |
| Mo/TiW | 3.60 E6 | >24,000 |

SOURCE: © 1987 IEEE.

resistance increase due to the large variation in the data. This data suggests that the addition of impurities to Al improves EM performance of the Al conductors. Even though the addition of Si to pure Al increases the MTTF by about three times, it is not very effective in combination with Cu addition. The MTTF of Al-Cu alloy is higher than the MTTF of the Al-Cu-Si alloy in this case.

Copper segregates at the grain boundaries. This slows the grain boundary diffusion process and increases the activation energy barrier. As a result, Al ions have to overcome a larger energy barrier to migrate. Copper concentrations ranging from 0.5 to 4 percent have been widely used in both bipolar and CMOS devices to enhance the EM resistance of interconnects. However, increasing the Cu concentration causes plasma etching (lack of volatile Cu species) and corrosion problems. Figure 4.24 shows a TEM image of the grains in an Al-Cu alloy film. The dark dots are the Cu atoms that have been segregated at the grain boundaries. As the temperature of metal deposition increases, the Cu segregation is enhanced. This can cause problems during plasma etching due to variation in the composition across the material.

**Impurities.** Impurities added during the metal deposition process degrade the quality of the metal film, that is, they alter the film properties. Vacuum integrity during sputter deposition of metal has to be very good and the residual gases monitored. Residual gases such as O, H, and N have a detrimental effect on deposition film.[26] Incorporation of $O_2$ in the film alters its resistivity and changes the film hardness. Excess $H_2$ incorporation leads to the formation of hillocks in Al films.[27] Hillock formation, as discussed in Chapter 3, may lead to yield and reliability problems.

Traces of nitrogen incorporation in the metal film have been found to severely degrade EM performance of the Al interconnect.[28] Figure 4.25 compares the relative EM performance of test samples with and without $N_2$ contamination and illustrates the impact of $N_2$ on EM performance. Nitrogen appears to contaminate the grain boundaries by

**Figure 4.24** Grain boundary distribution of Al-Cu film.

the formation of nitrides. This prevents grain growth during metal anneal operation and also stops motion along grain boundaries. $N_2$ contamination has also been found to change the stress of the metal film from tensile to compressive.[29]

**Grain size.** Grain boundary diffusion is one of the primary paths for movement of atomic flux in polycrystalline conductors. Diffusion of ions takes place via grains boundaries as they offer a much lower impedance. In single crystalline conductors, the atomic flux movement is through the lattice which results in a higher resistance to EM. Inhomogeneities in the microstructure can cause flux divergences that may result in EM failures only where grain boundaries contact dielectric films. The microstructure can be altered and degrade EM performance if the grains are misoriented and contaminated. Grain size variations can also cause inhomogeneities in the microstructure.

Electromigration failures in polycrystalline Al result in accumulation and depletion of material along the grain boundaries. As the grain size increases and exceeds the interconnection width, the microstructure approaches a "bamboo" structure where the availability of grain boundary diffusion paths is reduced. The more important effect is the reduction of number of grain boundaries which contact the dielectric. The bamboo structure creates a noncontinguous grain boundary system, as shown in Figure 4.26. Consequently, the EM resistance increases for a given current density, and approaches values for single crystalline material.

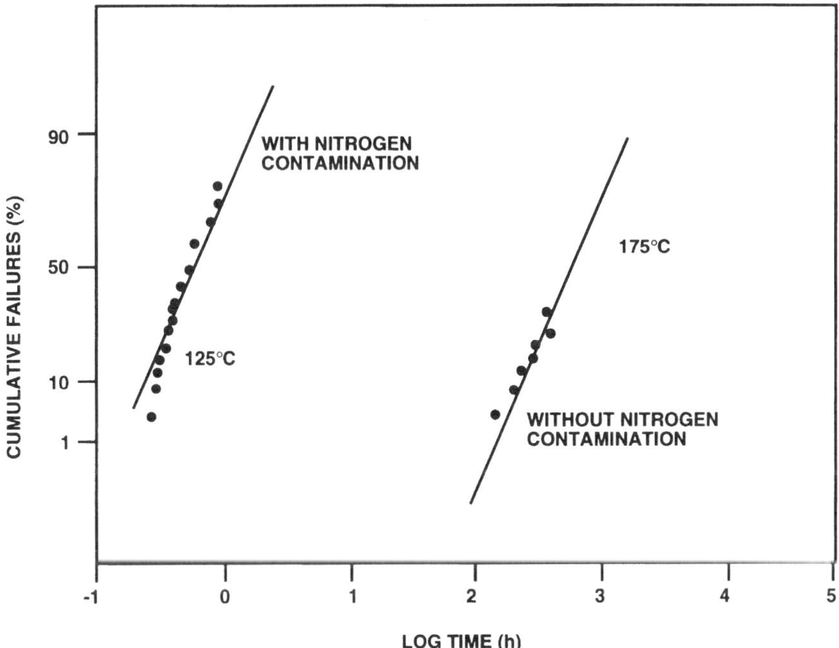


**Figure 4.25** Effect of nitrogen contamination on electromigration in Al.[28] (© *1984 IEEE.*)

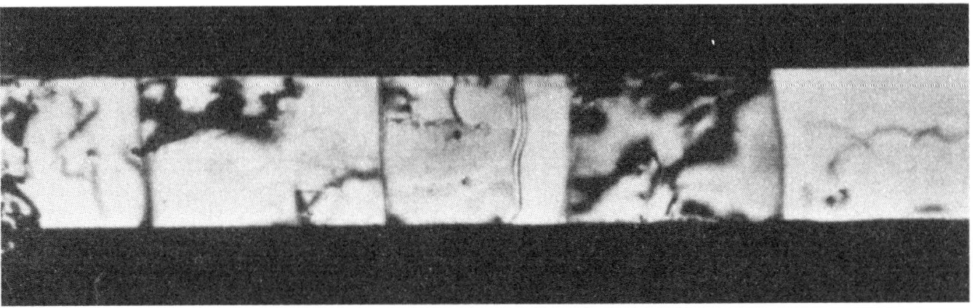

**Figure 4.26** Bamboo-type grain structure in narrow Al lines.

**Line length.** Multilevel interconnects contain lines with different lengths, widths, thicknesses, and spacing. Line length is varied and total length in a multilevel interconnect system may be as long as several meters. Line length impacts the EM performance of the conductor. It has been reported that the MTTF decreases as the length of the interconnect increases until a saturation point is reached. Figure 4.27 shows the impact on MTTF with increasing line length.[37] The standard deviation of the failure distribution is dependent on line length, but does not vary systematically with it. The initial decrease in MTTF and the subsequent saturation suggests that the defects that cause EM failures are randomly distributed and microstructure-dependent.

**Line width.** As line width becomes smaller the MTTF decreases. This process is shown in Figure 4.28. Although not shown in this diagram, the failure rate starts to increase at smaller line widths. The line width is also dependent on thickness and microstructure. As the line widths become smaller and smaller, the grain size of the Al interconnect approaches the bamboo structure, unless the underlying barrier metals limit the grain size. This microstructure enhances EM resistance, therefore MTTF increases. Even though MTTF increases, it has been found that the associated standard deviation increases with decreasing line width.[36]

There is another concern with decreasing metal line widths—stress-induced void formation. Stress migration failures increase as line

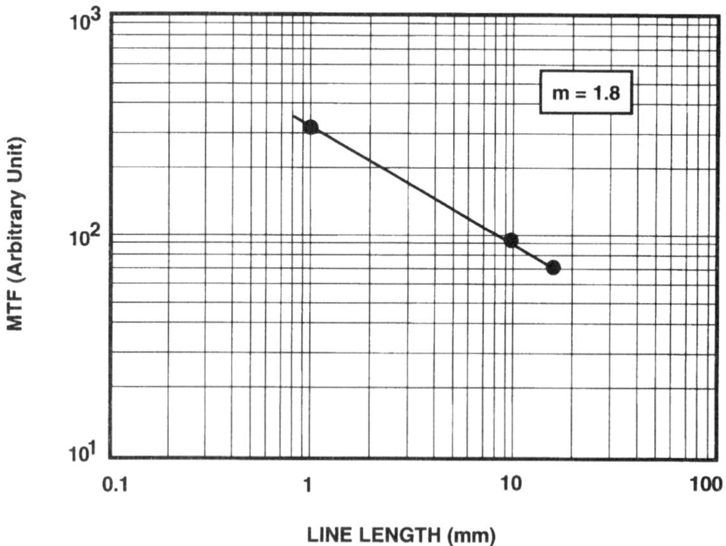


**Figure 4.27** Impact on mean time to failure due to variation in interconnect line length.[37] (© *1992 IEEE.*)

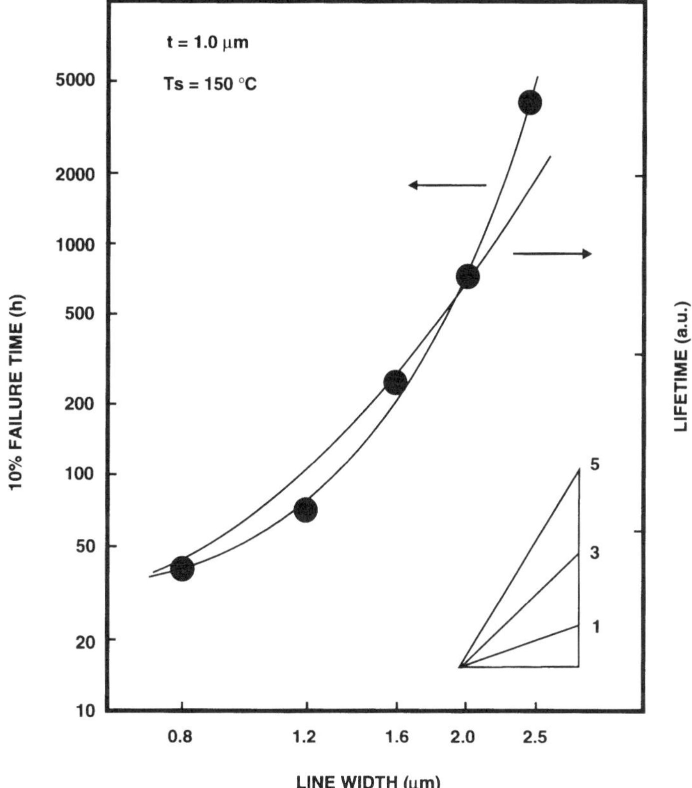


**Figure 4.28** Impact on mean time to failure due to variation in interconnect line width.[36] (© *1991 IEEE.*)

width decreases below approximately 3.0 $\mu$m. A typical range for interconnect width on current very large-scale integration (VLSI) circuits is about 1 to 2 $\mu$m, and this is shrinking as the integration level goes up. For the new generation of dynamic random access memories (DRAMs) and microprocessors, some of the interconnect line widths are in the submicron range. Stress migration can produce both large and narrow slit-like voids that result in open circuit failures. A more detailed analysis of void formation is presented in Section 4.45, below.

**Thickness.** In multilevel interconnect systems, the scaling results in thinning of the interconnects to make them compatible with the dielectric deposition over very tight pitches. It is generally considered that MTTF is not dependent on thickness for a constant current density.

**Step coverage.** Step coverage is a critical parameter that becomes prominent in multilevel interconnect systems. To improve the step coverage, a gamut of steps ranging from metal deposition to planarization

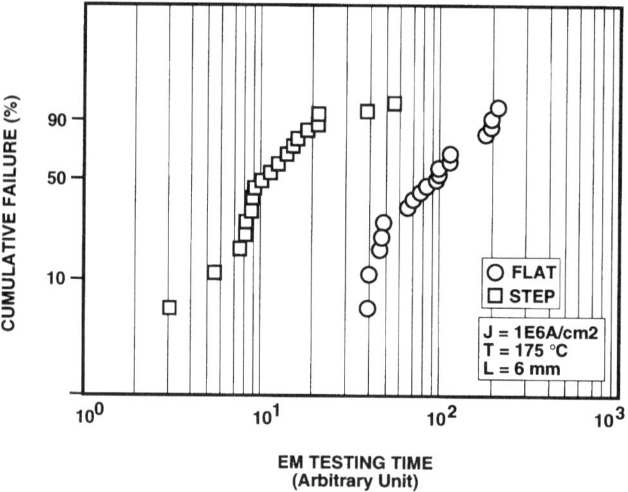


**Figure 4.29** Failure rate variation due to interconnect definition over steps.[37] (© *1992 IEEE.*)

are optimized for a particular product design. Poor metal step coverage has a detrimental effect on the EM performance of the interconnects. Degradation in EM in areas of poor step coverage is due to an increase in current density at steps as a result of metal thinning and smaller grain size. Electromigration degradation is typically much smaller over flat surfaces than over steps. Figure 4.29 shows the difference in failure rate with and without steps.[37] This figure illustrates that the failure time is much shorter for metal lines over a step.

**Stress/passivation.** Passivation significantly improves the EM performance of the conductor by exerting pressure on the metal lines. This pressure confines the metal and reduces material accumulation. Reduction of material accumulation in turn opposes EM. The passivation commonly used is silicon nitride or oxynitride as these provide a good hermetic barrier against moisture. Moisture penetration may lead to corrosion. Even though passivation improves EM resistance of the conductor, stress in passivation film may induce voids in Al lines. Stress effects and EM benefits need to be carefully managed. Internal stresses in Al films, the dielectric, and the passivation can all contribute to device failures. For example, thermal mismatch between Al films and the substrate can lead to formation of hillocks, which can cause shorts and degrade the yield and the reliability of the devices.

**Barrier metals.** In order to improve the overall reliability of the interconnects, barrier metals are used as part of the metal stack. A common material, titanium (Ti), is used as it provides an alternate current path wherever the Al has been made discontinuous by void formation. Barrier metals degrade the quality of the films by making its grains smaller, but provide adequate compensation by allowing a redundant current path. One of the problems with a Ti barrier layer is its interaction with Al. To prevent this, an Al-Si alloy must be used or another barrier layer between the Ti and the Al is needed such as TiN.

### 4.5.2 Contact/via electromigration

One of the most critical aspects of multilevel interconnect technology is the formation of contacts and vias and filling them with metal. The contact/via scenario has changed with the migration to submicron geometries where high-aspect-ratio contacts/vias with tight dimension control are the norm rather than the exception in circuit design. Figure 4.30 shows some of the major contact/via EM problems that were observed at different technology levels. For devices with minimum feature sizes larger than 1 $\mu$m, the dominant failure mechanisms were attributed to composition of Al alloys. Junction spiking was a major problem that was addressed by saturating the Al with Si and also by using a barrier layer. The side effect of this was the problem of Si precipitation in the contacts. This precipitation caused an increase in contact resistance and also degraded contact reliability. Step coverage of metal in contacts and vias is still a major issue. Contact sidewall tapering has allowed metal to be deposited in the contacts/vias with an adequate step coverage.

The current submicron problems with contacts and vias are different. The density of contacts/vias has increased tremendously and the design rules for contact size, placement, and layout have been tightened. With the migration to submicron features, the aspect ratio of contacts and vias has increased. Consequently, step coverage of the metal interconnect is a major issue. This has been overcome to a great extent by the use of a contact filling processes. In this regard, W

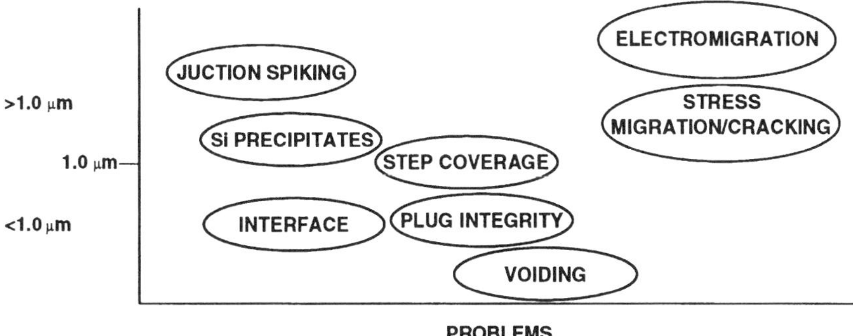


**Figure 4.30** Some common interconnect reliability problems at different process technology generations.

plugs and studs have found wide application. There are still many problems associated with the contact metallization process that degrade reliability of the devices. Perhaps the key element to ensuring a high level of contact/via EM performance lies in the plug integrity. Voids in and near the plug may cause poor metal step coverage of the interconnect metal. Other problems may be due to the interface of the plug and the interconnect.

The via interface is a major cause for concern, both in terms of via resistance and EM. The interaction of contact metal with the underlying metal interconnect at this interface must be carefully evaluated. One of the common problems during via etch is controlling the overetch and penetration into the underlying metal interconnect. This is partially due to the varying topography. As described in Chapter 2, the Al alloy conductor may be layered with other films to make it more robust. Overetching the vias may create a situation where material interactions at the interface dominate performance. For example, one of the commonly used metals for filling contacts and vias is W. In a recent study,[34] EM performance between W and Al contact/via fill materials was compared. It is interesting to note that Al-Al vias exhibited better EM performance than Al-W vias. Having a diffusion barrier layer at the via interface improves EM performance as it suppresses or reduces the void formation.[32,33] Interface cleaning is essential after the aggressive via etch. Sputter etch using Ar is also widely used to clean the bottom of contacts and vias. It has been shown[31,32] that sputter etch improves EM performance. Let us summarize the salient issues with via EM, as given in Table 4.10.

### 4.5.3 Voiding

Voiding in Al and its alloys is a serious problem, especially when critical dimensions are shrinking below the submicron marker. *Voids* are basically areas of depleted or missing material in the conductor, as

**TABLE 4.10 Via EM Issues**

| Issue | Effects |
|---|---|
| Overetch | Interface damage, polymer formation, exposure of Al underlying layer |
| Sputter etch | Improves contact/via resistance and also EM performance |
| Metal stack | Improves line EM, provides an etch stop during via etch |
| Barrier layer | At via interface, prevents void formation |
| Aspect ratio | Increase in aspect ratio degrades EM performance due to current crowding |

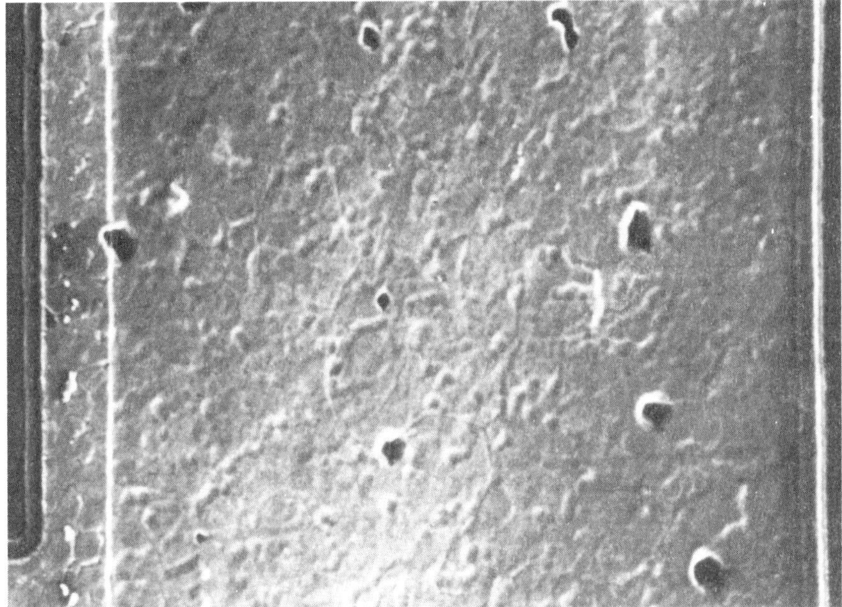

**Figure 4.31** Stress-induced voiding in Al conductors.

shown in Figure 4.31. These voids tend to increase in size and propagate through the conductor. Over time, these voids can grow large enough to break the continuity of the metal and cause an open circuit failure. Hence it is important to understand the parameters that modulate void growth and examine ways to reduce or even eliminate their formation. There are several factors that can cause voiding to take place. To examine these, a cause and effect diagram is used as shown in Figure 4.32. In addition, Table 4.11 gives the impact of various parameters on void formation. The arrows indicate the direction of change. It is evident that void formation is not just stress-induced, but is the product of several factors that interact with one another. Two types of voids may be observed on metal interconnects. These are the slit-type and wedge-type voids. Initially, voids that are formed at the grain boundaries take on a wedge-shape appearance that deforms into a slit shape.[48]

Metal lines underneath passivation are in a state of triaxial tensile stress. Stress in the metal is built up during the postpassivation temperature processing. When Al cools down, it experiences a volume shrinkage that is more than the passivation shrinkage. This volume reduction causes an increase in stress in the metal film. This volume reduction is further aggravated when a layered metallization scheme is employed, and intermetallic reactions can also lead to some volume

**TABLE 4.11 Factors Impacting Void Formation**

| Parameter | Change | Voids | Remarks |
|---|---|---|---|
| Grain size | ↑ | ↑ | Size and distribution determine void growth and propagation |
| Line width | ↓ | ↑ | Narrow metal |
| Thickness | ↓ | ↑ | Metal thickness |
| Stress | ↑ | ↑ | Thermal mismatch between Al and silicon substrate. Dielectric layer acts as a constraint. |
| Passivation | ↑ | ↑ | Passivation stress induces voids; thickness/stress factors |
| Temperature | ↑ | ↑ | Accelerates void growth |
| Time | ↑ | ↑ | Incubation time allows |
| EM effects | ↑ | ↑ | Current stress |

shrinkage. The passivation exerts a hydrostatic stress on the metal underneath. Stress relief is accomplished in the metal by the formation of voids that start at grain boundaries that contact the dielectric. The dielectric acts as a constraint and keeps the metal from shrinking so it is left in hydrostatic tension, that is, it is metastable. It could be that defects on the metal plus other catalyzing factors trigger the void. Narrow metal lines may create severe conditions because of their surface area/volume ratio. Further growth and propagation of the voids is dictated by the amount of stress in the metal film and the possibility of atomic migration paths.

## 4.6 Summary

Ensuring high reliability of on-chip and package level multilevel interconnect systems is extremely important as this reflects the success of the entire IC design, fabrication, and assembly processes. The focus of this book has been on the issues associated with on-chip interconnect systems since extensive research and development is being done in this field to enhance the performance of advanced ICs. Specifically, the issues discussed relate to IC design, process, and high-volume manufacturing. The impact of these issues on yield and reliability have been examined and discussed. From this anslysis it is clear than an integrated approach needs to be taken to gain a comprehensive understanding of multilevel interconnect technology and its successful application.

## References

1. D. J. Klinger et al. (eds.), *AT&T Reliability Manual,* Van Nostrand Reinhold, 1990.
2. S. M. Sze (ed.), *VLSI Technology,* 2d ed., McGraw-Hill, 1988.
3. *Components Quality and Reliability,* Intel, 1991–1992.
4. E. R. Hnatek, *Integrated Circuit Quality and Reliability,* Marcel Dekker, 1987.

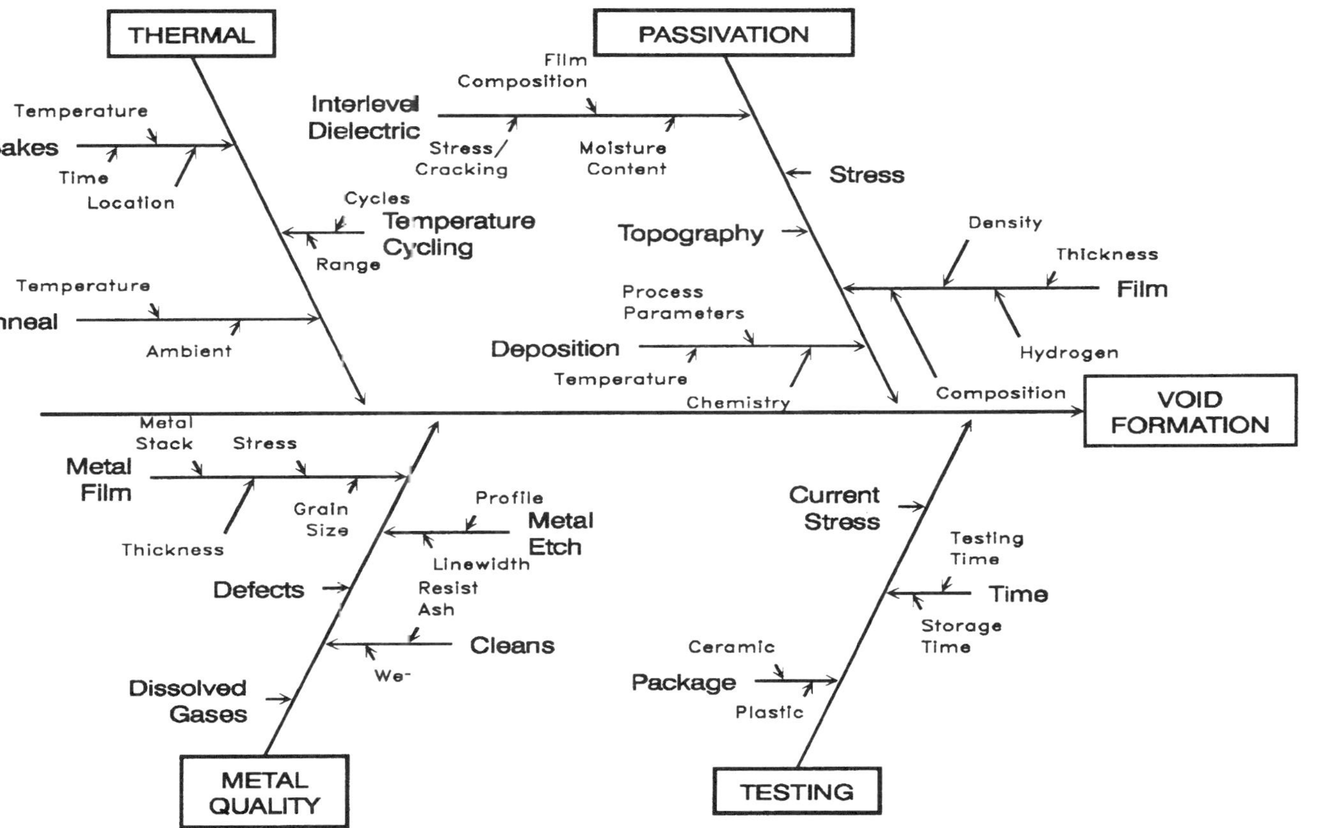


Figure 4.32 Cause and effect diagram for void formation.

5. H. B. Bakoglu, *Circuits, Interconnections and Packaging for VLSI,* Addision-Wesley, 1990.
6. D. L. Crook, "Evolution in VLSI reliability engineering," *IRPS,* 1990.
7. P. E. Fieler, "Understanding Motorola's six sigma program," *Tutorial Notes, IRPS,* 1990.
8. H. H. Huston and C. P. Clarke, "Reliability defect detection and screening during processing—theory and implementation," *IRPS*, 1992.
9. D. A. Baglee and D. S. Gibson, "Wafer-level reliability implementation issues," *Tutorial Notes, IRPS,* 1990.
10. H. A. Schafft et al., "Building-in reliability: Making it work," *IRPS,* 1991.
11. S. Wolf, *Silicon Processing for the VLSI Era,* Vol. 2, *Process Integration,* Lattice Press, 1990.
12. S. W. Mittl and M. J. Hargrove, "Hot carrier effects in P-channel MOSFETS," *IRPS,* 1989.
13. C. C. H. Hsu, "Hot-carrier induced instability of 0.5 $\mu$m CMOS devices patterned using synchrotron x-ray lithography," *IRPS,* 1989.
14. A. Q. Testone, *Static Electricity in the Electronics Industry,* Testone Enterprises, 1990.
15. O. J. McAteer, *Electrostatic Discharge Control,* McGraw-Hill, 1989.
16. X. Guggenmos, "MOS degradation in input and output stages of VLSI CMOS circuits due to electrostatic discharge," *Journal de Physique,* Colloque C4, Supplement No. 9, Vol. 49, September 1988.
17. D. Zupac et al., "Detection of ESD-induced catastrophic damage in P-channel power MOSFETS," *EOS / ESD Symp. Proc.,* 1991.
18. B. A. Unger, "Electrostatic discharge failures of semiconductor devices," *IRPS,* 1981.
19. R. Pepe, "ESD repeatability in the real world," *Evaluation Engineering,* July 1990.
20. Y. Fong and C. Hu, "Internal ESD transients in input protection circuits," *IRPS,* 1989.
21. *Proc. of the IRPS,* 1983–1992.
22. J. R. Lloyd et al. (eds.), *Proc. of the Symp. on Electromigration of Metals and First Intl. Symp on Multilevel Metallization and Packaging, ECS,* Vol. 85-6, 1985.
23. H. S. Rathore et al., *Proc. of the Symp. on Reliability of Semiconductor Devices and Interconnection and Multilevel Metallization, Interconnection and Contact Technologies, ECS,* Vol. 89-6, 1989.
24. N. G. Einspruch et al. (eds.), *VLSI Electronics Microstructure Science,* Vol. 15, Academic Press, 1987.
25. P. B. Ghate, "Current reliability issues in multilevel interconnection", *State-of-the-Art Seminar Visuals Booklet,* Course Coordinator T. E. Wade, 1987.
26. A. R. Nyaiesh and L. Holland, "Effects of gas composition on discharge and deposition characteristics when magnetron sputtering aluminum," *Vacuum,* 31, Nos. 8/9, 1981.
27. D. R. Denison, "Sputtering system design for optimum deposited film quality," *Microelectronics Manufacturing and Testing,* July 1985.
28. J. Klema et al., "Reliability implications of $N_2$ contamination during deposition of sputtered Al/Si films," *IRPS,* 1984.
29. L. Holland, *Vacuum Deposition of Thin Film,* Chapman and Hall, London, 1970.
30. *Proc. VMIC,* 1987–1992.
31. P. Freiberger and K. Wu, "A novel via failure mechanism in Al-Cu/Ti double level metal system," *IRPS,* 1992.
32. N. D. Bui et al., "Effect of barrier metal, of grain size, and of interface cleanliness on electromigration performance of via chain," *VMIC,* 1882.
33. T. Yamaha et al., "Three kinds of via electromigration failure models in multilevel interconnections," *IRPS,* 1992.
34. J. Tao et al., "Comparison of electromigration reliability of tungsten and aluminum vias under DC and time varying current stressing," *IRPS,* 1992.
35. D. Basile et al., "FIBXTEM—focussed ion beam milling for TEM sample preparation," *Mat. Res. Soc. Symp. Proc.,* Vol. 254, 1992.

36. T. Hosoda et al., "Effects of line size on thermal stress in aluminum conductors," *IRPS*, 1991.
37. T. Nogami et al., "Electromigration lifetime as a function of line length and step number," *IRPS*, 1992.
38. I. Gaeta and K. J. Wu, "Improved EPROM moisture performance using spin-on-glass (SOG) for passivation," *IRPS*, 1989.
39. C. C-H. Hsu et al., "Hot-carrier induced instability of 0.5 $\mu$m CMOS devices patterned using synchrotron x-ray lithography," *IRPS*, 1989.
40. V. Jain et al., "Internal passivation for suppression of device instabilities induced by backend processing," *IRPS*, 1992.
41. J. Takahashi et al., "Water trapping effect of point defects in interlayer plasma CVD $SiO_2$ Films," *VMIC*, 1992.
42. B. van Schravendijk et al., "Correlation between dielectric reliability and compositional characteristics of PECVD oxide films," *VMIC*, 1992.
43. R. C. Blish and P. R. Vaney, "Fail rate model for thin film cracking in plastic ICs," *IRPS*, 1991.
44. R. L. Zelenka, "A reliability model for interlayer dielectric cracking during temperature cycling," *IRPS*, 1991.
45. C. G. Shirley and R. C. Blish, "Thin film cracking and wire ball shear in plastic DIPs due to temperature cycle and thermal shock," *IRPS*, 1987.
46. T. M. Moore et al., "Improving plastic package reliability," *Tutorial Notes, IRPS*, 1992.
47. M. J. Kim et al., "Electromigration of bias sputtered Al and comparison with others," *IRPS*, 1987.
48. H. Kaneko et al., "A newly developed model for stress induced slit-like voiding," *IRPS*, 1990.
49. N. Hirashita et al., "Effects of residual water in SOG layer on void formation for multilevel interconnections," *IRPS*, 1990.
50. Y. Ohji et al., "The effect of minute impurities (H, OH, F) on $SiO_2$/Si interface investigated by nuclear resonant reaction and electron spin resonance," *IRPS*, 1989.
51. C. Jiang et al., "Impact of intermetal oxide deposition condition on NMOS and PMOS transistors hot carrier effects," *IRPS*, 1992.

Appendix

# Overview of Cause and Effect Interactions

## A.1 Overview

Cause and effect diagrams have been used throughout this book to examine all the pertinent issues relating to a process step or problem. Additional cause and effect diagrams are included in this Appendix to present an overview of multilevel interconnect technology in its entirety and the problems encountered with it. It is revealing to explore the interaction between a process technology step and problems that may arise due to process marginalities or variations. Table A.1 is a matrix comparing common problems associated with multilevel interconnect technology with various individual components of process technology. Two kinds of interactions are listed. First is technology-technology interaction, which is observed when moving down the technology column of the table. Second is technology-problem interaction, which is evident when moving laterally.

The cause and effect diagrams presented in this book provide an extensive network going from interconnect design to packaging issues. Taken by themselves, these diagrams provide an excellent tool in understanding individual technology issues. When combined with cause and effect diagrams for problems, they provide an even more effective tool for solving problems quickly. This intricate web of cause and effect diagrams presents the reader with a broad range of issues and concerns associated with multilevel interconnect technology.

**TABLE A.1 Overview of Technology—Problems and Cause and Effect Interactions**

| | | Problems | | | | | | | | | |
|---|---|---|---|---|---|---|---|---|---|---|---|
| Technology | Figure no. | Contact resistance | Step coverage | Corrosion | Hot carrier | EM | ESD | Stress voids | Yield | Particles | Thin-film cracking |
| Figure no. | | 2.18 | A.4 | A.9 | 4.8 | 4.23 | 4.12 | 4.32 | 3.16 | 3.35 | 4.19 |
| Design | A.1 | √ | √ | — | √ | √ | √ | — | — | — | √ |
| Silicides | A.2 | √ | — | — | — | — | — | — | √ | √ | — |
| Dielectric | 2.3 | √ | √ | √ | √ | √ | — | √ | √ | √ | √ |
| Planarization | | √ | √ | — | — | √ | — | — | √ | √ | — |
| CMP | 2.9 | | | | | | | | | | |
| SOG | 2.13 | | | | | | | | | | |
| Lithography | A.3 | √ | √ | — | — | — | — | — | √ | √ | — |
| Etch | | | | | | | | | | | |
| Contact | 2.16 | √ | √ | — | √ | √ | — | — | √ | √ | — |
| Metal | A.5 | √ | √ | √ | √ | √ | — | — | √ | √ | — |
| Cleans | A.6 | √ | — | √ | — | — | — | — | √ | √ | — |
| Metal | | | | | | | | | | | |
| W plug | 2.24 | √ | √ | √ | √ | √ | — | — | √ | √ | — |
| Deposition | 2.28 | √ | √ | √ | √ | √ | — | √ | √ | √ | √ |
| Line definition | 2.27 | √ | √ | √ | √ | √ | — | √ | √ | √ | √ |
| Passivation | A.7 | — | — | √ | √ | √ | — | √ | √ | √ | √ |
| Packaging | A.8 | — | — | √ | — | — | √ | √ | — | — | √ |

√ = strong interaction
— = weak/no interaction.
*Abbreviations:* CMP = chemical mechanical planarization; EM = electromigration; ESD = electrostatic discharge; SOG = spin-on-glass; W = tungsten.

## A.2 Technology/Problem Cause and Effect Analysis

### A.2.1 Design

Figure A.1 puts into perspective various issues that are of concern when designing circuits with multilevel interconnects. Interconnect design impacts not only device performance but also how processes are integrated together. The design issues in this diagram have been discussed in detail in Chapter 1. Design changes affect every aspect of downstream technology. For example, dielectric deposition can be severely influenced by metal spacing. As the designs are introduced with smaller geometries, pressure to develop more effective process technologies increases. As we move horizontally across the Design row of Table A.1, several problems arise due to tight design tolerances or circuit layout schemes conducive to yield (parametric and/or functional) or reliability degradation.

### A.2.2 Silicides

Use of refractory metal silicides to lower source/drain and polysilicon interconnect resistance has become the mainstay for advanced CMOS integrated circuits. Refractory metals are reacted with silicon to form a silicide layer. A maskless, self-aligned process of forming silicide layers is achieved as only areas with silicon exposed react to form refractory metal silicide. The reaction is initiated under high temperature, either in a furnace or in single-wafer rapid thermal processing equipment. Various factors that influence refractory metal silicide formation are presented in Figure A.2. Since the silicide layer is an etch stop during contact etch, any damage due to overetching can contribute to an increase in contact resistance. A key issue with silicide films is their interaction with downstream processing. For example, the interface between silicide film and the tungsten plug adhesion layer must be understood since a good portion of the contact resistance may be attributed to this interface. In addition to contributing to contact resistance, the silicide process may also contribute to yield loss by poor or incomplete formation, agglomeration, and other related defects. Factors that can vary silicide formation are surface cleanliness and implant species. High temperatures after silicide formation could degrade the silicide quality due to grain boundary oxidation.

### A.2.3 Lithography

Along with multilevel interconnect technology, lithography is a key process technology that impacts design and fabrication of sub-half-micron devices. The usable limit of optical lithography is fast

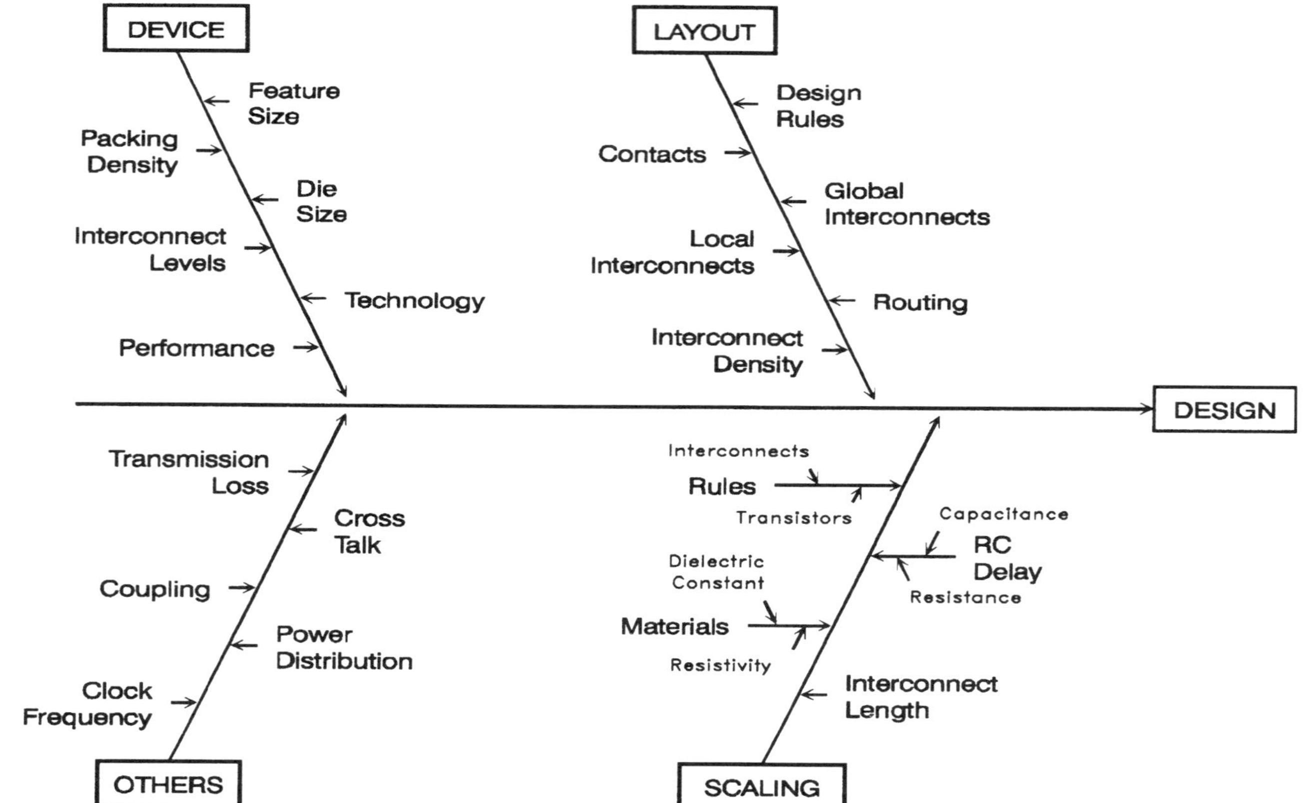


**Figure A.1** Cause and effect diagram for interconnect design issues.

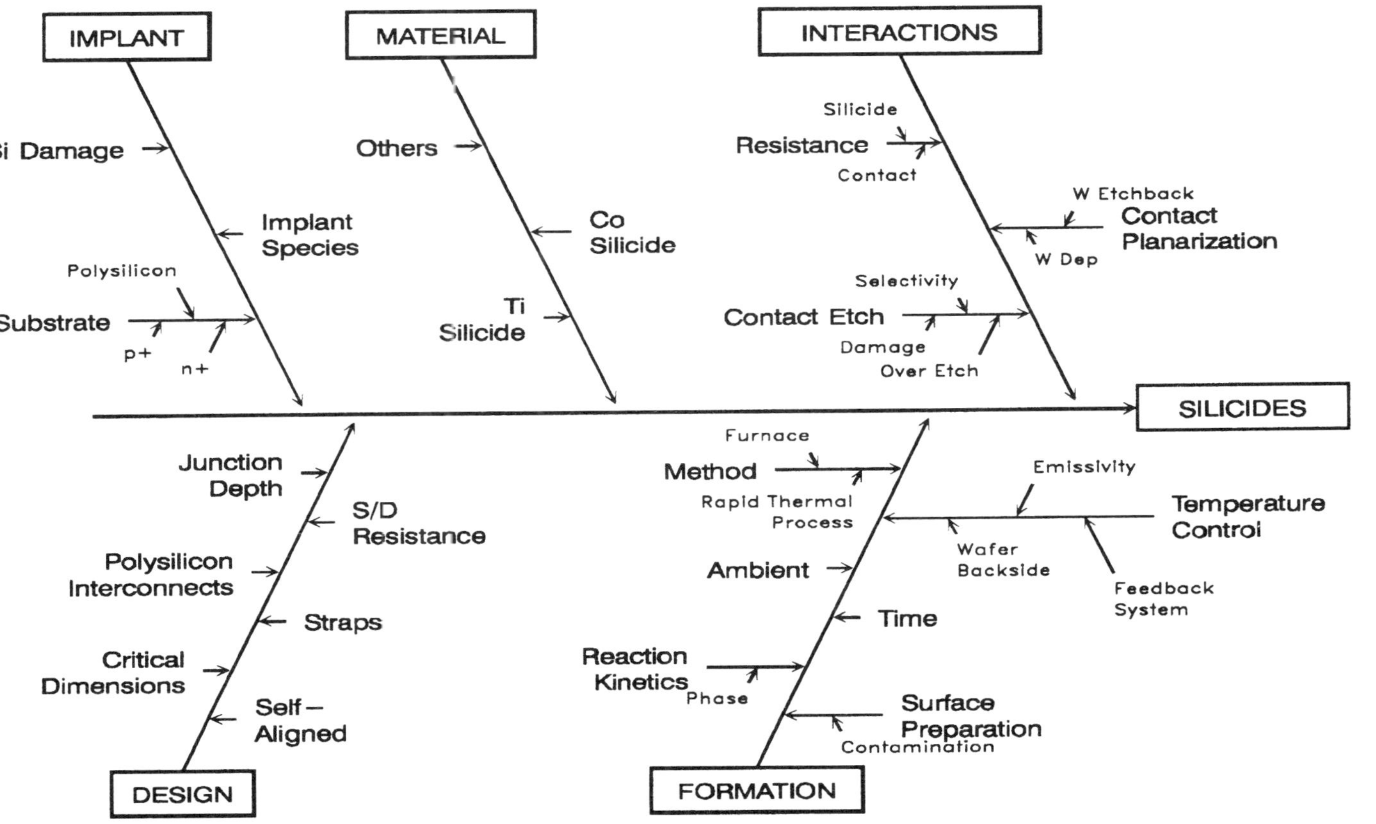


**Figure A.2** Cause and effect diagram for silicides.

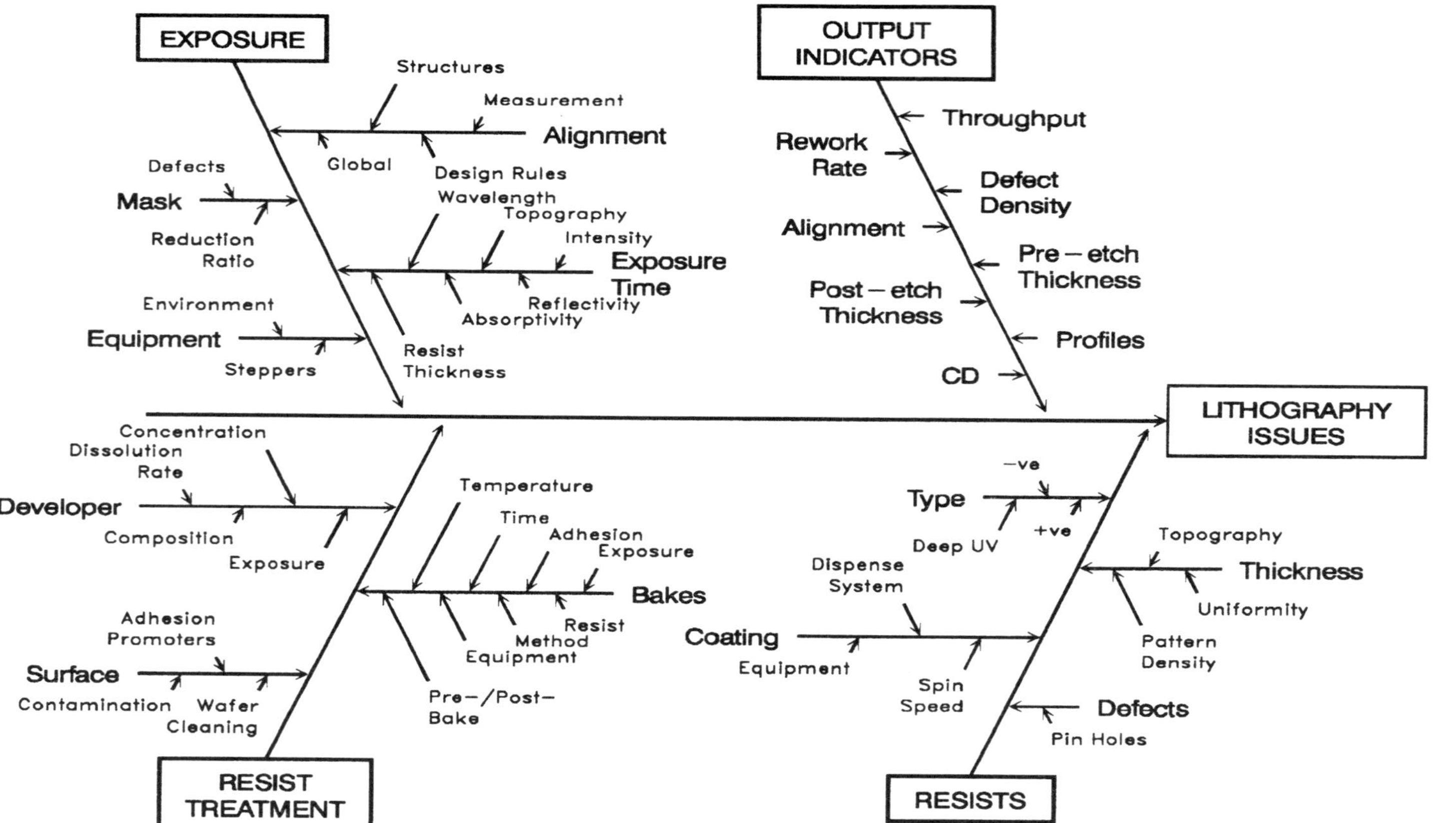


**Figure A.3** Cause and effect diagram for lithography issues.

approaching and x-ray, electron-beam, etc., devices are in various stages of development as possible replacements. Figure A.3 examines the salient aspects of optical lithography. Lithography impacts etch processes and can contribute to several problems. These problems can range from high contact resistance (due to low critical dimensions) to a systematic yield loss (due to defects in the reticles). The selection and characterization of resists and developers for a given process technology have a tremendous impact on the performance of etch processes. Resist coating, exposure, and development have to be controlled in order to minimize rework rate. Among the output parameters that the lithography area is responsible for are critical dimensions, alignment, rework rate, etc. However, to achieve these output parameters satisfactorily, each of the component processes must be optimized for each layer due to variations in substrates and planarity.

### A.2.4 Dielectric deposition

In Chapter 2, two kinds of dielectric films were discussed: those that are used before and after metallization steps. As these films are used for isolation between conducting layers, the impact on device yield and reliability is significant. Adverse process variations and interactions with downstream processing can contribute to various problems, as shown in Table A.1.

### A.2.5 Planarization

Several planarization methods are currently in use. Each has its advantages and disadvantages. Key issues here are metal step coverage and fine pattern printing. As the number of metal layers increases, metal step coverage becomes a major concern. Figure A.4 shows different factors that can vary metal step coverage. Poor metal step coverage can alter both current density and the electromigration performance of devices. Besides planarization, metal deposition and circuit design also play a critical role in determining metal step coverage. Step coverage over contacts and vias is of course also modulated by contact and plug processes.

### A.2.6 Etch

In the multilevel interconnect process flow there are two critical etch steps. These are contact/via and metal etch. The first step is a challenge since very small contacts must be opened in planarized dielectric films, resulting in contacts of different depths. All these contacts must be etched with appropriate selectivities to resist and substrate materials. Contact etch is one of the major modulators of contact resistance. The second critical etch is metal etch; Figure A.5 shows

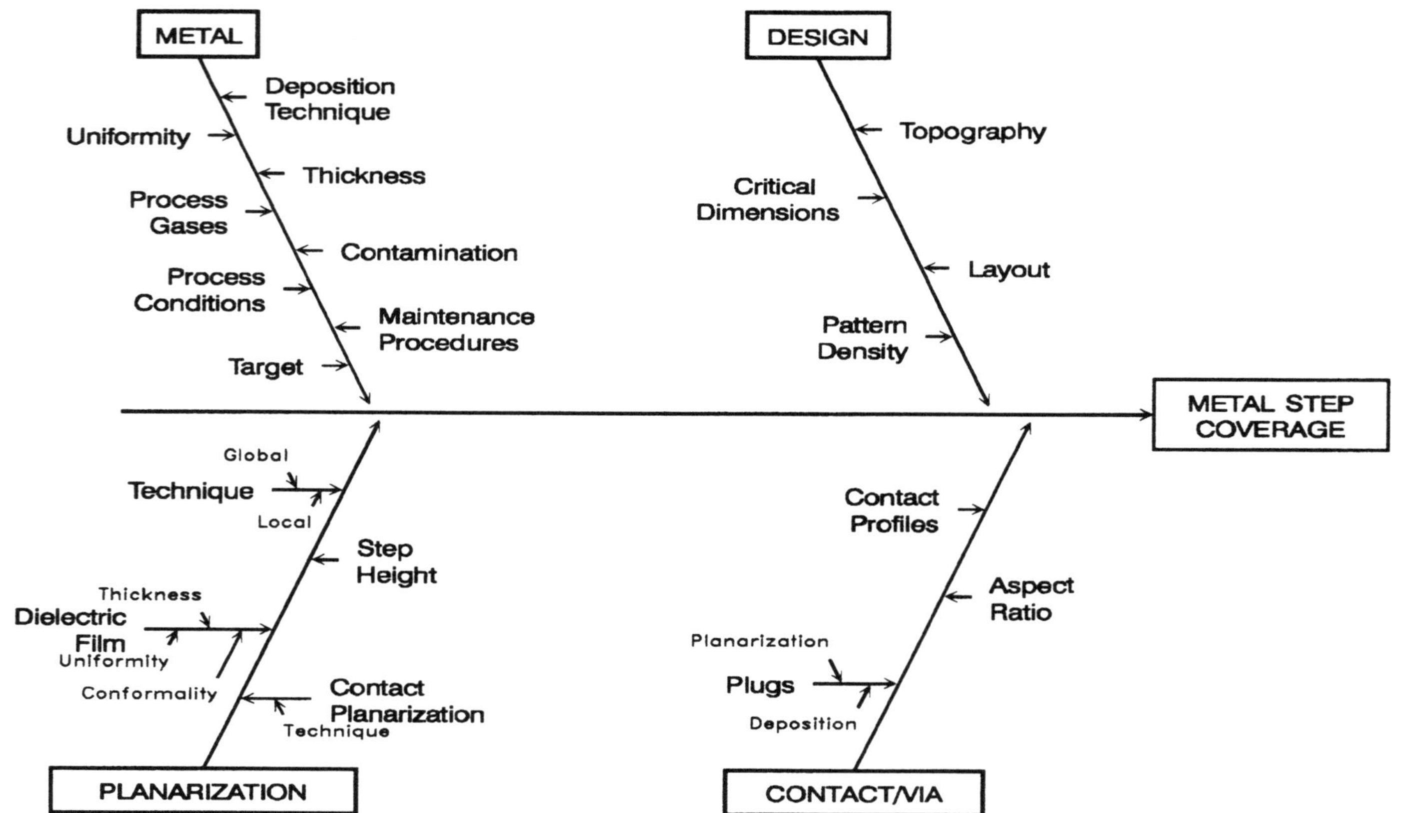


**Figure A.4** Cause and effect diagram for metal step coverage.

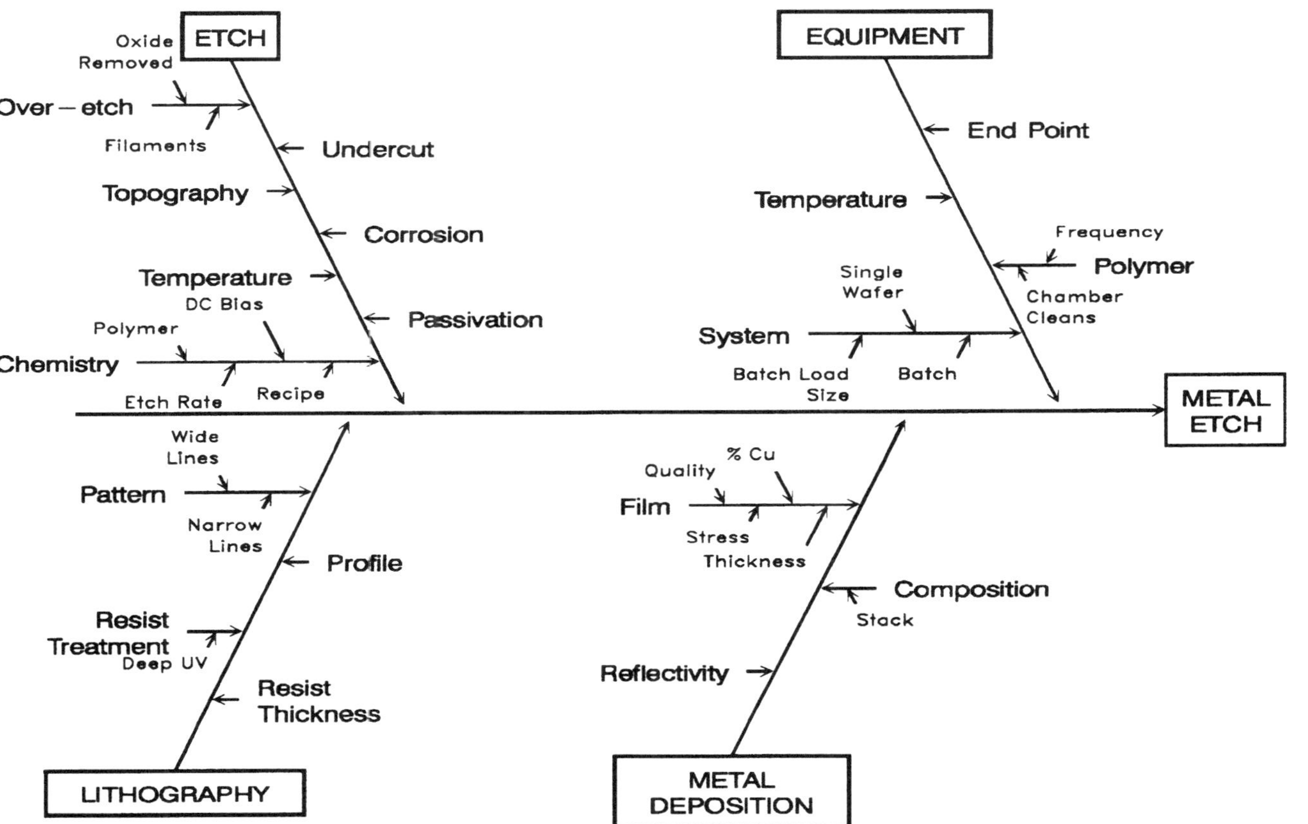


**Figure A.5** Cause and effect diagram for metal etch.

**TABLE A.2 Etch Selectivities**

| Contact etch | Via etch | Blanket W etchback | Metal etch | Planarization | Passivation etch |
|---|---|---|---|---|---|
| BPSG: resist | Oxide: resist | W: adhesion layer | Al alloy: resist | Resist/SOG: oxide | Nitride: resist |
| >> 1 | >> 1 | 1:1 | >> 1 | 1:1 | >> 1 |
| BPSG: silicide | Oxide: metal stack | W: oxide | Al alloy: barrier/ARC layer | — | — |
| >>>> 1 | >> 1 | > 1 | 1:1 | | |
| BPSG: Si (n + /p +) | — | — | Al alloy: shunt layer | — | — |
| >>> 1 | | | 1:1 | | |
| BPSG: polysilicon | — | — | Al alloy: oxide | — | — |
| >>> 1 | | | > 1 | | |

*Abbreviations:* Al = aluminum; ARC = antireflective; BPSG = borophosphosilicate glass; Si = silicon; SOG = spin-on-glass; W = tungsten.
> = greater than; >> = significantly greater than; >>> =very significantly greater than.

the cause and effect diagram for metal etch. A major challenge for metal etch is control and management of corrosion. Careful examination of a typical metal etch process reveals that several metals are being etched at the same time.

*Selectivity,* the ratio of etch rates between two films that are being etched simultaneously, is one of the critical etch parameters. Table A.2 compares selectivities of various etch processes. The selectivity of a film may be expressed as follows:

$$S = ER_A/ER_B \tag{A.1}$$

where $S$ = selectivity
$ER$ = etch rate
A = thin film A
B = thin film B

### A.2.7 Cleans

Resist removal is an integral part of any etch or implant process where resist is used as a mask (see Figure A.6). During etching, resist masks are exposed to ion bombardment and temperature. This changes the texture of resist, and under certain conditions it can lead to reticulation and hardening. In the case of metal etch, the resist

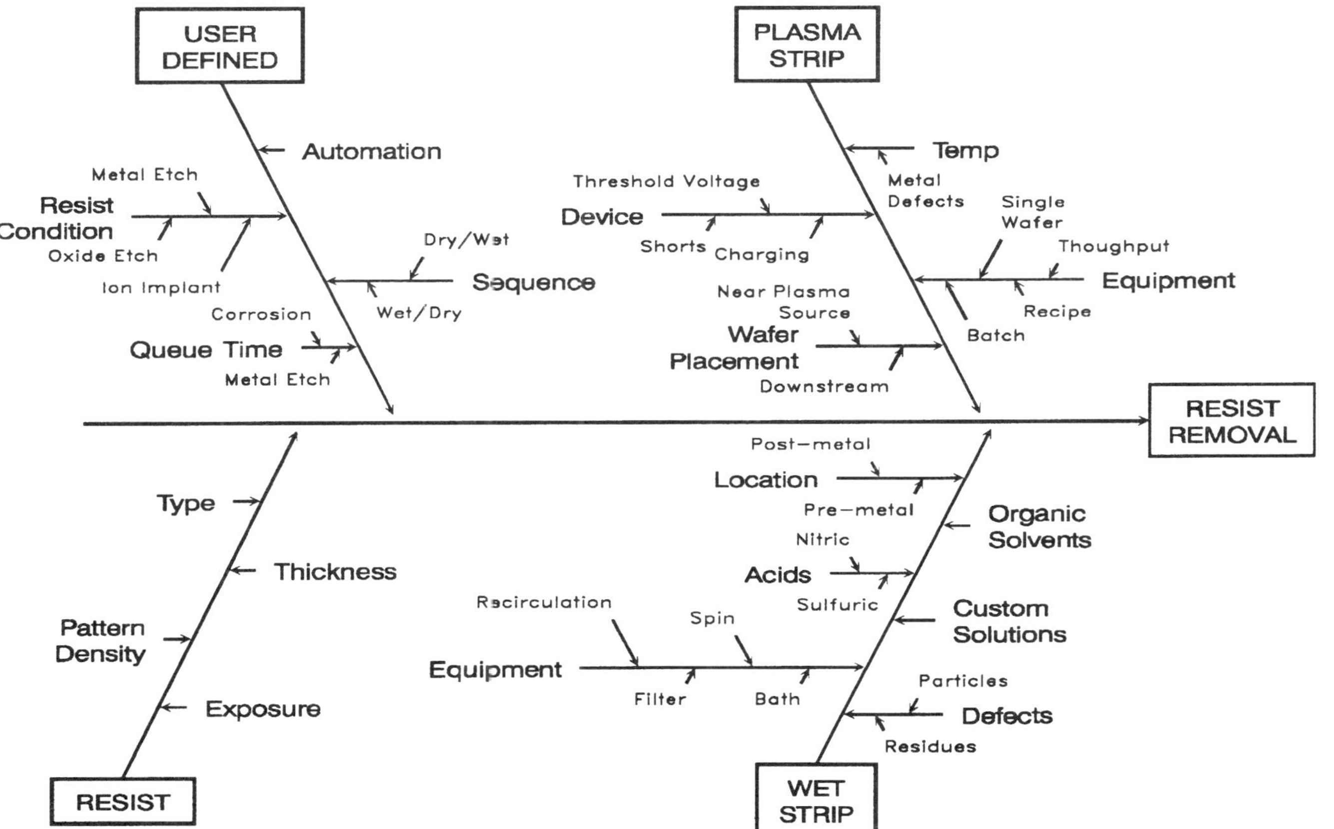


**Figure A.6** Cause and effect diagram for resist removal.

absorbs a significant amount of chlorine that causes corrosion on exposure to air. Removing hardened resist and chlorine are two of the challenges faced during resist removal. Resist may be removed using plasma systems or by wet chemical means. Usually a combination of both techniques is employed to make the resist removal operations more effective. Plasma ashing systems can also cause device instability. To overcome this problem, ashers with a plasma source that is situated away from the wafers are widely used.

### A.2.8 Metallization

Metallization has a significant effect on device functionality and reliability. Quality of metal films, corrosion, step coverage, and etching characteristics are some of the major issues. Several cause and effect diagrams in this book examine the interactions of metallization from various perspectives, for example, contact resistance modulation and stress voiding.

### A.2.9 Passivation

Depositing a protective overcoat of silicon nitride or other films is a prerequisite for ensuring that the fully processed die can withstand assembly process steps and field conditions. Passivation film protects the die from scratches and moisture. However, some problems are associated with the use of passivation films such as silicon nitride. The stress in this film can cause voids to form in metal interconnects. In addition, passivation films interact with the packaging material, and depending on stresses imposed on the die, thin film cracking may occur. Figure A.7 shows several factors that can modulate passivation integrity. Integrity of the film is so important that even minute pinholes in the film can provide a path for moisture to penetrate and cause corrosion of the underlying metal.

### A.2.10 Packaging

Packaging technology is yet another expanding field that utilizes multiple levels of interconnects. Although a detailed discussion of packaging issues is beyond the scope of this book, it is essential to understand how packaging processes and materials can effect the die. For example, a thin-film cracking problem is influenced by the stress induced by the package. Another example of how packaging may affect device reliability may be observed by monitoring corrosion. Figure A.8 presents an overview of package-induced corrosion. Assembly process and handling are two of the causes of corrosion in metal interconnects (see Figure A.9). Of course, the die is only as good

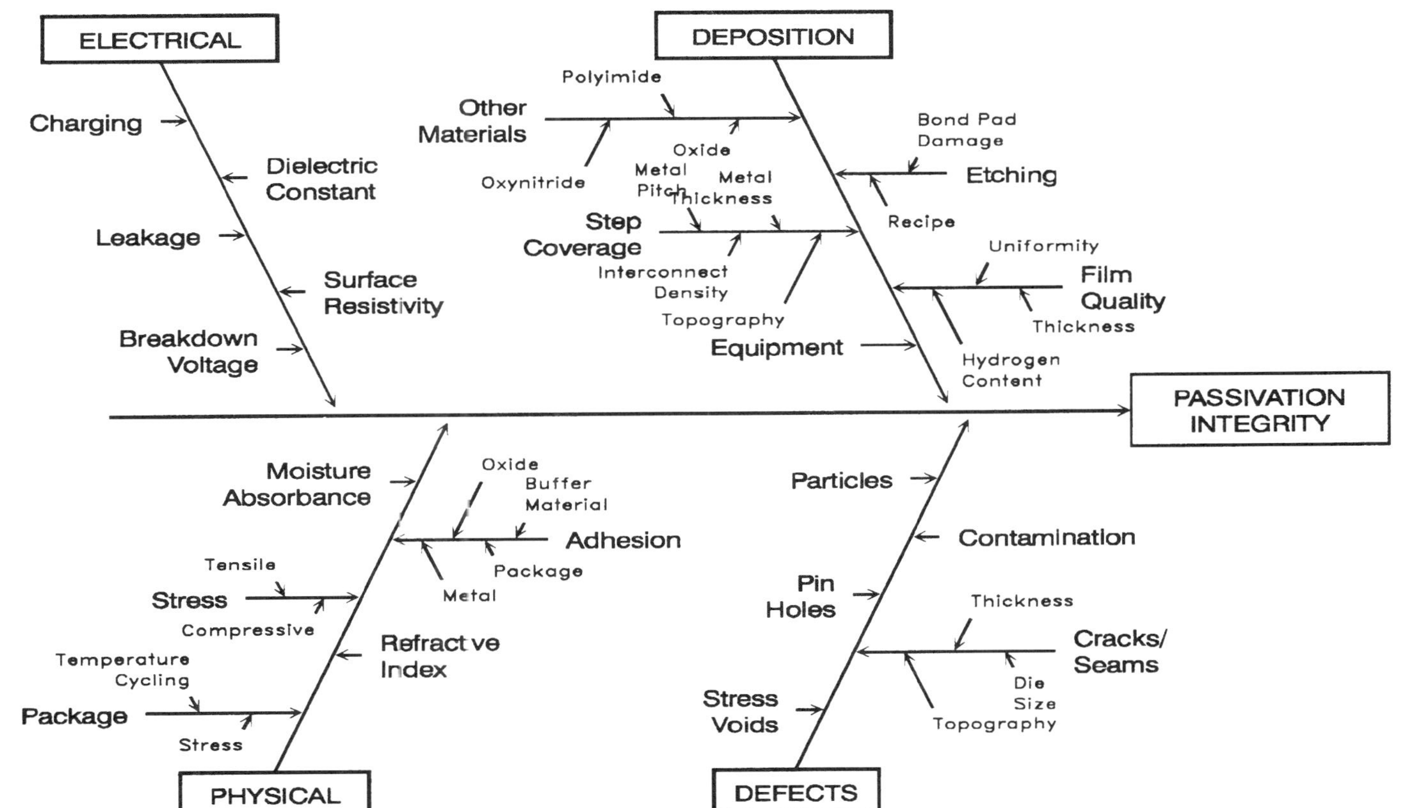


Figure A.7 Cause and effect diagram for passivation integrity.

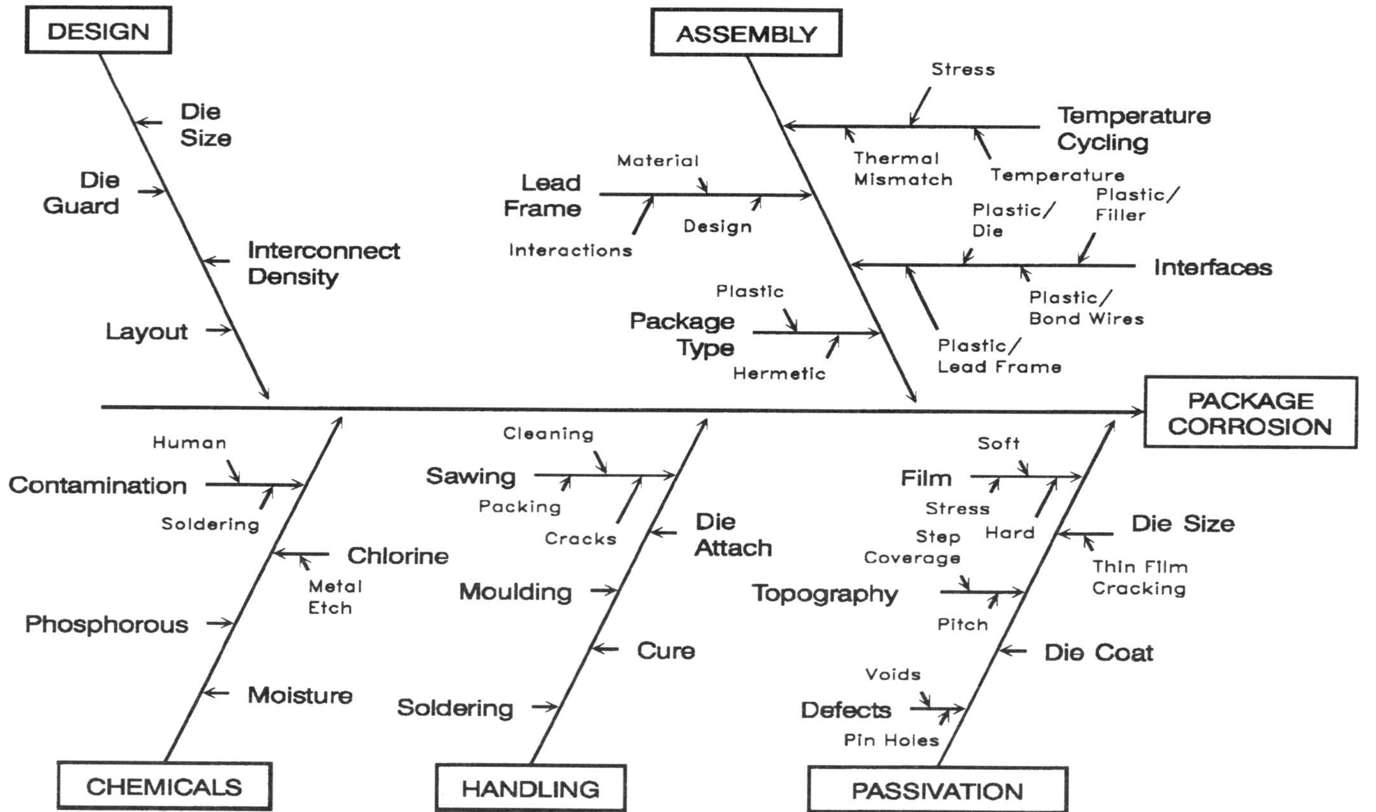


**Figure A.8** Cause and effect diagram for plastic package corrosion.

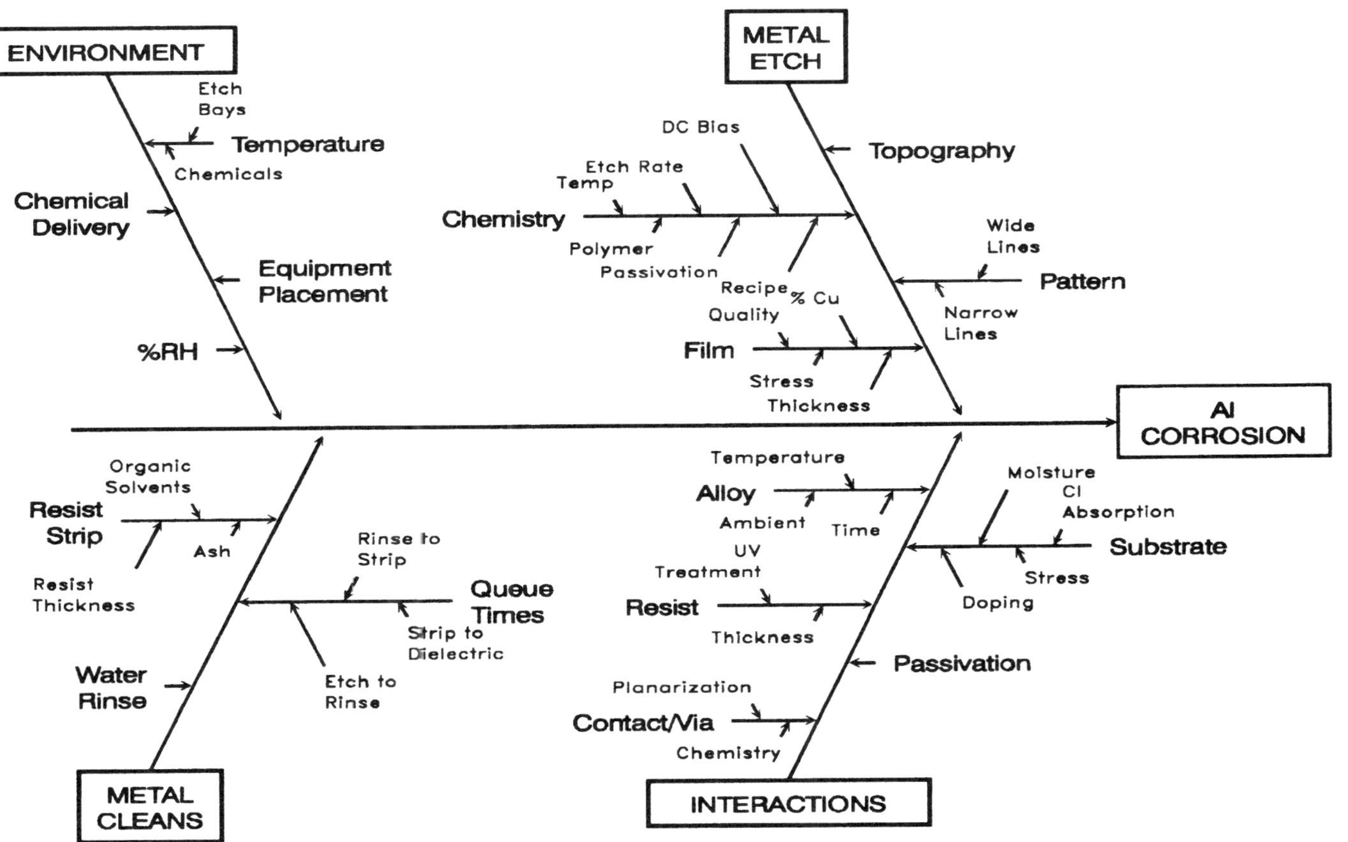


**Figure A.9** Cause and effect diagram for Al interconnect corrosion.

**TABLE A.3 Process Equipment Particles Contamination Levels**

| | 1-μm particles added | |
|---|---|---|
| Process equipment | Total | Particles/cm$^2$ |
| Plasma via etch | 1300 | 17 |
| Plasma oxide etch | 1000 | 13 |
| Oxide etch hood | 890 | 11 |
| Resist asher | 420 | 5 |
| Drying oven | 290 | 4 |
| Epitaxy | 280 | 4 |
| Ion implanter | 250 | 3 |
| Multiprobe | 185 | 2 |
| LPCVD oxide | 160 | 2 |
| Metal sputter | 150 | 2 |
| Plasma silicon etch | 140 | 2 |
| Oxide furnace tube | 100 | 1 |
| Passivation deposition | 85 | 1 |
| Current goal | 16 | 0.2 |
| Future need (0.1 μm and larger) | 0.8 | 0.001 |

LPCVD = low-pressure CVD.
SOURCE: Courtesy Dr. B. Y. H. Liu, University of Minnesota.

as its passivation layer and the quality of the fabrication process steps.

### A.2.11 Particles

Particles are a common problem across the board with any process step. Monitoring particle levels in chemicals, equipment, and processes is a routine task in any fabrication plant. Reducing these levels is the charter for defect reduction groups as particles can cause yield degradation. Table A.3 compares the particle levels for different sets of equipment. Such a comparison is useful in determining the level of effort required to reduce particle levels in equipment. The strategy for reduction varies from one type of equipment to another. This requires yet another cause and effect type of analysis, taking into account all the hardware and software aspects of the equipment.

# Index